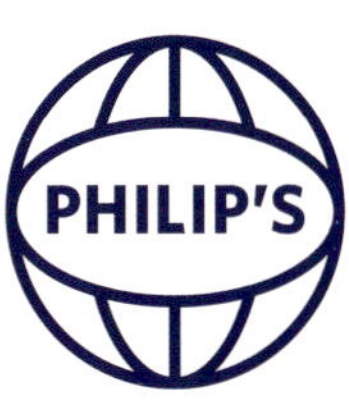

BRITAIN & IRELAND

NIGEL HENBEST

STARGAZING 2026

MONTH-BY-MONTH GUIDE TO THE NIGHT SKY

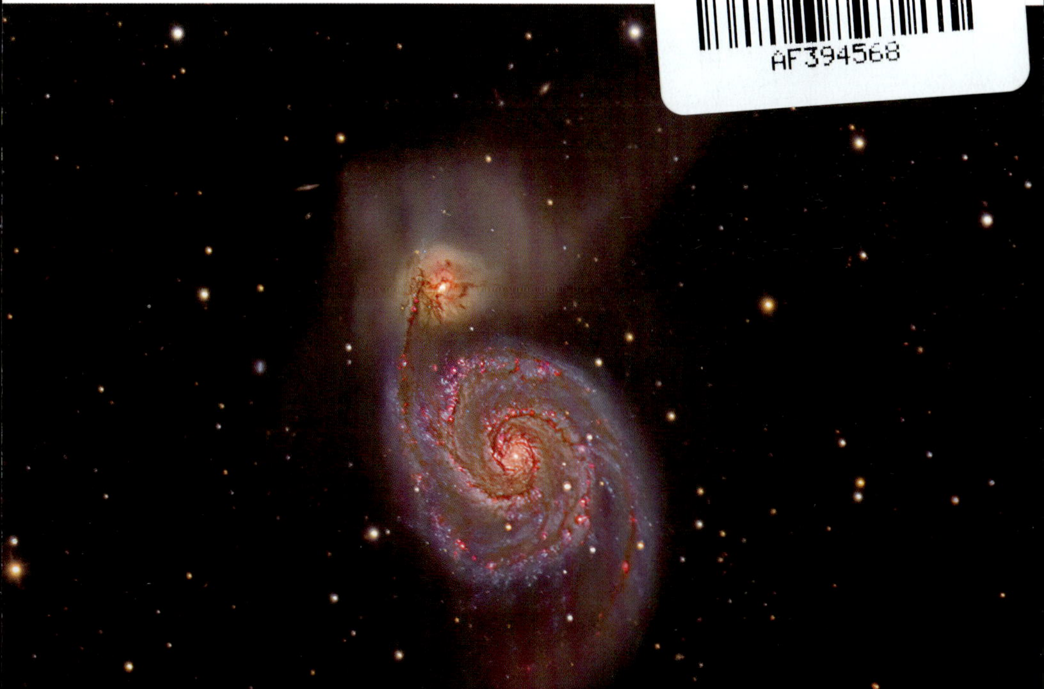

www.philips-maps.co.uk

First published in Great Britain in 2025
by Philip's, an imprint of
Octopus Publishing Group Ltd
Carmelite House
50 Victoria Embankment
London EC4Y 0DZ
www.octopusbooks.co.uk

An Hachette UK Company
www.hachette.co.uk

The authorized representative in the EEA is Hachette
Ireland, 8 Castlecourt Centre, Dublin 15, D15 XTP3, Ireland
(email: info@hbgi.ie)

TEXT
Nigel Henbest © 2025 pp. 4–85, 90–95
Philip's © 2025 pp. 1–3
Robin Scagell © 2025 pp. 86–89

ARTWORKS © Philip's except pp. 13, 19, 25, 31, 37, 43, 49,
55, 61, 67, 73, 79 © Nigel Henbest/Philip's

ISBN 978-1-84907-725-5
eISBN 978-1-84907-726-2

A CIP catalogue record for this book is available from the
British Library.

Printed in China
10 9 8 7 6 5 4 3 2 1

Cover: Aurora borealis over Stonehenge, 5 May 2024.
Title page: M51, the Whirlpool Galaxy.

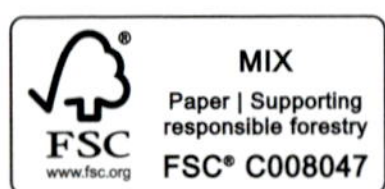

CONTENTS

Welcome to the latest edition of *Stargazing*! Within these pages, you'll find a complete guide to everything happening in the night sky throughout 2026 – whether you're a beginner or an experienced astronomer.

With the 12 monthly Star Charts, you can find your way around the sky on any night in the year. Impress your friends by identifying celestial sights ranging from the brightest planets to some pretty obscure constellations.

Every page of *Stargazing 2026* is bang up-to-date, bringing you everything that's new this year, from shooting stars to eclipses. And I'll start with a run-down of the most exciting sky sights on view in 2026 (opposite).

THE STAR CHARTS

A reliable map is just as essential for exploring the heavens as it is for visiting a foreign country. So each monthly section starts with a circular **Star Chart** showing the whole evening sky.

To keep the maps uncluttered, I've plotted about 200 of the brighter stars (down to third magnitude), which means you can pick out the main star patterns – the constellations. (If the charts showed every star visible on a really dark night, there'd be around 3000 stars on each!) I also show the ecliptic: the apparent path of the Sun in the sky; it's closely followed by the Moon and planets as well.

You can use these charts throughout the UK and Ireland, along with most of Europe, North America and northern Asia – between 40 and 60 degrees north – though the precise timings apply specifically to Britain and Ireland.

USING THE STAR CHARTS

It's pretty easy to use the charts. Start by working out your compass points. South is where the Sun is highest in the sky during the day; east is roughly where the Sun rises, and west where it sets. At night, you can find north by locating the Pole Star – Polaris – by using the stars of the Plough (see March's Object).

The left-hand chart then shows your view to the north. Most of the stars here are visible all year: these circumpolar constellations wheel around Polaris as the seasons progress. Your view to the south appears in the right-hand chart; it changes much more as the Earth orbits the Sun. Leo's prominent 'Sickle' is high in the spring skies. Summer is dominated by bright Vega, Deneb and Altair. Autumn's familiar marker is the Square of Pegasus; while the stars of Orion rule the winter sky.

During the night, our perspective on the sky also alters as the Earth spins round, making the stars and planets appear to rise in the east and set in the west. The charts depict the sky in the late evening (the exact times are noted in the captions). As a rule of thumb, if you are observing two hours later, then the following month's chart will be a better guide to the stars on view – though beware: the Moon and planets won't be in the right place.

THE PLANETS, MOON AND SPECIAL EVENTS

The charts also highlight the **planets**

HIGHLIGHTS OF THE YEAR

- **9 January:** Jupiter is closest to the Earth.
- **27 January:** the Moon occults the Pleiades.
- **17 February:** an annular eclipse of the Sun is visible from Antarctica.
- **18 February:** the crescent Moon lies between Venus and Mercury.
- **19 February:** Mercury's greatest separation from the Sun sees its best evening appearance.
- **20 February:** Saturn passes close to Neptune.
- **3 March:** a total eclipse of the Moon is visible from the Americas, Asia and Australasia, but not from Ireland or the UK.
- **7 March:** Venus passes close to Neptune.
- **8 March:** Venus near Saturn.
- **20 March:** the crescent Moon and Venus form a stunning duo.
- **29 March:** Regulus occulted by the Moon.
- **19 April:** Venus and the Moon form a beautiful tableau near the Pleiades.
- **Night of 22/23 April:** the Lyrid meteor shower is at its best in the early morning sky.
- **23 April:** Venus passes Uranus.
- **18 May:** the crescent Moon is close to Venus.
- **9 June:** conjunction of Venus and Jupiter.
- **17 June:** Venus near the crescent Moon.
- **19 June:** Venus passes near Praesepe.
- **4 July, am:** Mars passes very close to Uranus.
- **17 July:** the crescent Moon is near Venus.
- **12 August:** Iceland and Spain experience a total eclipse of the Sun; Ireland and the UK a very large partial eclipse.
- **Night of 12/13 August:** maximum of the Perseid meteor shower.
- **15 August:** Venus, at its greatest evening separation from the Sun, is near the Moon.
- **28 August (am):** a large partial eclipse of the Moon is visible from the Americas, Africa and Europe, including the UK and Ireland.
- **8 September, am:** the Moon occults Praesepe.
- **9 September:** peak of the sparse 45P-id meteor shower.
- **14 September:** Venus lies close to the Moon.
- **22 September:** Venus reaches its greatest brilliance as the Evening Star.
- **25 September:** Neptune nearest to the Earth and opposite to the Sun.
- **4 October:** Saturn lies opposite the Sun in the sky and closest to Earth.
- **5 October, 6 am:** the Moon skims over Mars.
- **11 October, before dawn:** Mars appears in the middle of Praesepe.
- **Night of 21/22 October:** maximum of the Orionid meteor shower.
- **Night of 27/28 October:** the Moon occults the Pleiades.
- **7 November, am:** the crescent Moon teams up with Venus.
- **16 November:** Mars passes close to Jupiter.
- **Night of 17/18 November:** maximum of the Leonid meteor shower.
- **20 November, am:** Mercury reaches its greatest separation from the Sun on its best morning appearance of the year.
- **26 November:** Uranus is closest to the Earth.
- **27 November:** Venus at its greatest brilliance as the Morning Star.
- **28 November:** the Moon occults Praesepe.
- **5 December, am:** Venus is near the crescent Moon.
- **Nights of 13/14 and 14/15 December:** an excellent year for the Geminid meteor shower.
- **21 December:** the Moon occults the Pleiades.
- **24 December:** supermoon – the biggest and brightest Full Moon since 2018.

above the horizon in the late evening. I've indicated the track of any **comets** known at the time of writing; though I can't guide you to a comet that's found after the book has been printed!

I've plotted the position of the Full Moon each month, and also the **Moon's position** at three-day intervals before and afterwards. If there's a **meteor shower** in the month, the charts show its radiant – the position from which the meteors stream outwards.

The **Calendar** provides a daily guide to the Moon's phases and other celestial happenings. I've detailed the most interesting in the **Special Events** section, including close pairings of the planets, times of the equinoxes and solstices and

– most exciting – **eclipses** of the Moon and Sun.

Check out the **Planet Watch** page for more about the other worlds of the Solar System, including their antics at times they're not on the monthly Star Charts. I've illustrated unusual planetary and lunar goings-on in the **Special Event Charts**. There's a full guide to planetary motions, eclipses and meteor showers in **Solar System 2026** on pages 80–82.

MONTHLY OBJECTS, TOPICS AND PICTURES

Each monthly section highlights a fascinating **object** – a planet, a star or a nebula – as well as a stunning **picture** taken by an amateur based in Britain or Ireland along with full technical information. There's also an in-depth exploration of an intriguing and often newsworthy **topic**, ranging from blue moons to dark matter, and from Mayan astronomy to the student who discovered pulsars.

GETTING IN DEEP

A practical **Observing Tip** is included each month, helping you to explore the sky with the naked eye, binoculars or a telescope.

Check out my guide to the **Top 20 Sky Sights**, including nebulae, star clusters and galaxies. You'll find it on pages 83–85.

And equipment expert Robin Scagell provides his insights into the best telescopes now on the market for anyone just starting out in astronomy – and what to avoid! – on pages 86–89, whether you want to take photos or just eyeball the sky.

The final section details the internationally approved dark-sky sites in Britain and Ireland, where you're guaranteed to be free of light pollution. It also includes a round-up of planetariums and public observatories you can visit, plus a guide to the best star camps and astronomical festivals.

Happy stargazing!

Today, scientists can measure the light from the stars with amazing accuracy. (Mathematically speaking, a difference of five magnitudes represents a difference in brightness of one hundred times.) So the Pole Star is magnitude +2.0, while Rigel is magnitude +0.1. Because we've inherited the ancient ranking system, the brightest stars have the *smallest* magnitude. In fact, the most brilliant stars come in with a negative magnitude, including Sirius (magnitude –1.5).

And we can use the magnitude system to describe the brightness of other objects in the sky, such as stunning Venus, which can be almost as brilliant as magnitude –5. The Full Moon and the Sun have whopping negative magnitudes!

At the other end of the scale, stars, nebulae and galaxies with a magnitude fainter than +6.5 are too dim to be seen by the naked eye. Using ever larger telescopes – or by observing from above Earth's atmosphere – you can perceive fainter and fainter objects. The most distant galaxies visible to the Hubble Space Telescope are ten billion times fainter than the naked-eye limit.

Here's a guide to the magnitude of some interesting objects:

Sun	-26.7
Full Moon	-12.5
Venus (at its brightest)	-4.9
Sirius	-1.5
Betelgeuse (variable)	0.0 – +1.6
Polaris (Pole Star)	+2.0
Faintest star visible to the naked eye	+6.5
Faintest star visible to the Hubble Space Telescope	+31

Degrees of separation

Astronomers measure the distance between objects in the sky in **degrees** (symbol °): all around the horizon is 360°, while it's 90° from the horizon to the point directly overhead (the zenith).

As you can see in the photograph, it's possible to use your hand – held at arm's length – to give a rough idea of angular distances in the sky.

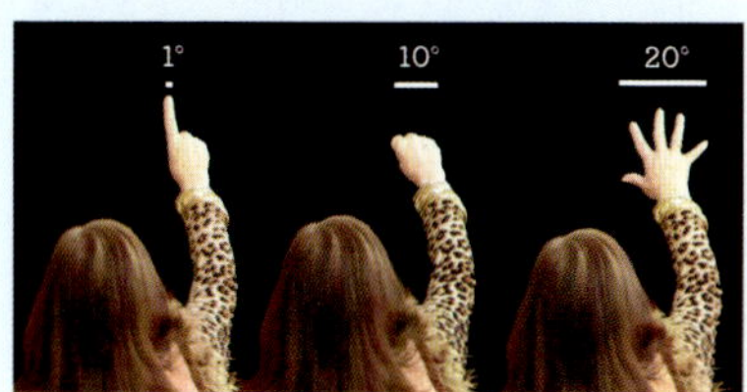

For objects that are very close together – like many double stars – we divide the degree into 60 arcminutes (symbol '). And for celestial objects that are extremely tiny – such as the discs of the planets – we split each arcminute into 60 arcseconds (symbol "). To give you an idea of how small these units are, it takes 3600 arcseconds to make up one degree.

Here are some typical separations and sizes in the sky:

Length of the Plough	25°
Width of Orion's Belt	3°
Diameter of the Moon	31'
Separation of Mizar and Alcor	12'
Diameter of Jupiter	45"
Separation of Albireo A and B	35"

How far's that star?

Everything we see in the heavens lies a long way off. We can give distances to the planets in millions of kilometres. But the stars are so distant that even the nearest, Proxima Centauri, lies some 40 million million kilometres away. To turn those distances into something more manageable, astronomers use a larger unit: one **light year** is the distance that light travels in a year.

One light year is about 9.46 million million kilometres. That makes Proxima Centauri a much more manageable 4.2 light years away from us. Here are the distances to some other familiar astronomical objects, in light years:

Sirius	8.6
Polaris	440
Centre of the Milky Way	26,700
Andromeda Galaxy	2.5 million
Most distant galaxies seen by James Webb Space Telescope	13.5 billion

- The sky at 10 pm in mid-January, with Moon positions at three-day intervals either side of Full Moon.
- The star positions are also correct for 11 pm at the beginning of January, and 9 pm at the end of the month.
- The planets move slightly relative to the stars during the month.

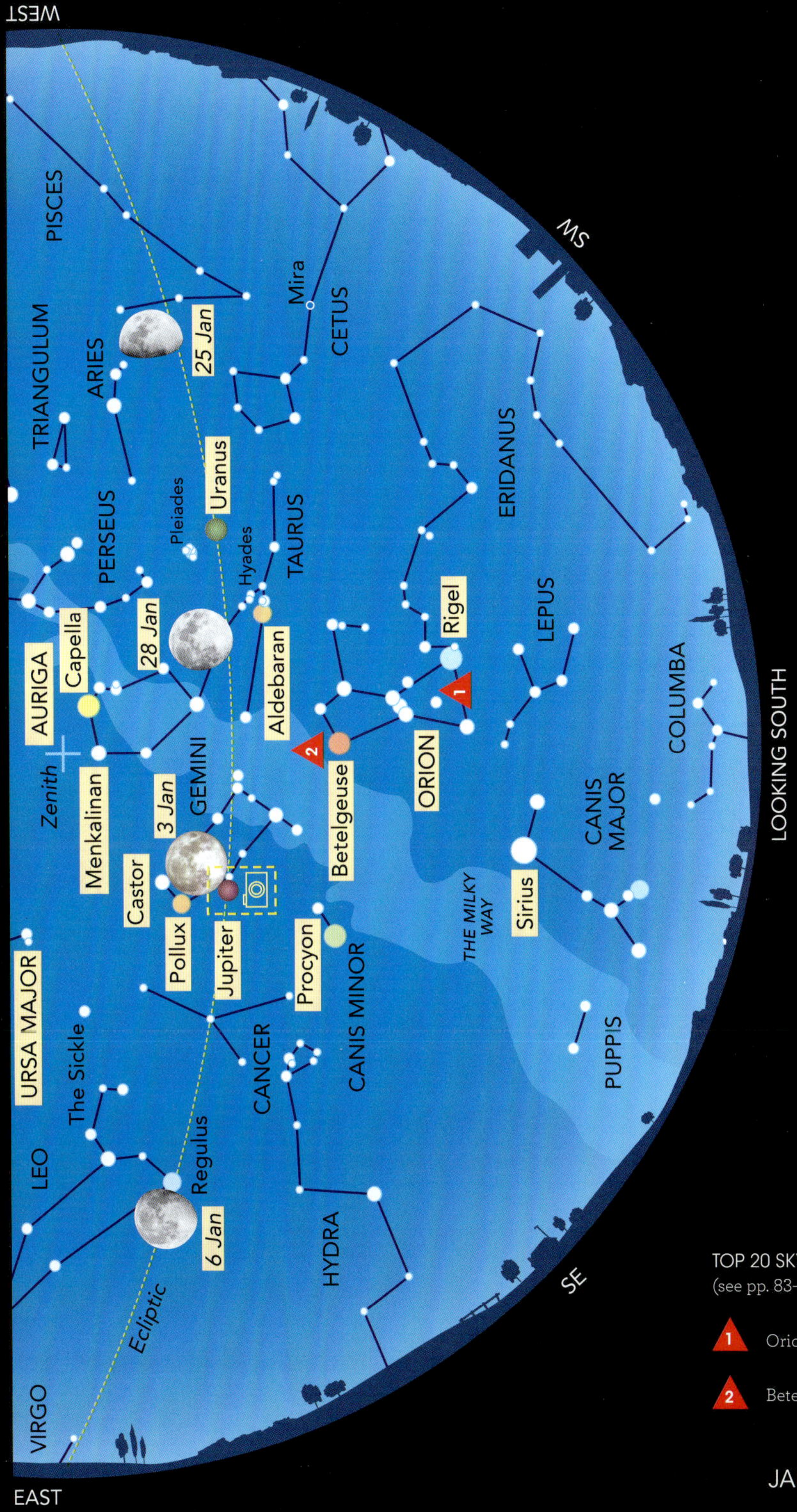

TOP 20 SKY SIGHTS
(see pp. 83–85)

1 Orion Nebula

2 Betelgeuse

The new year opens with a tableau of dazzling stars: **Betelgeuse** and **Rigel** in **Orion**, escorted by glorious **Sirius** and – forming a giant arc above – **Procyon**, **Castor** and **Pollux**, **Capella** and the red giant **Aldebaran**. The jewel in this scintillating crown is the brilliant planet **Jupiter**, at its closest to Earth this month.

JANUARY'S CONSTELLATION

Ursa Major, the Great Bear, is a constellation with deep roots. This star pattern may date back 30,000 years, when Palaeolithic people in Siberia erected shrines to cave bears. The bear tradition was then exported both to southern Europe and to the New World.

The seven brightest stars are often called the **Plough** or, in North America, the Big Dipper. You can use its two end stars – **Merak** and **Dubhe** – to locate **Polaris**, the Pole Star (see March's Object). In the middle of the bear's tail you'll find one of the few double stars that you can 'split' with the unaided eye: **Mizar** and its fainter companion **Alcor**. A small telescope reveals Mizar itself to be a close double.

Unlike most constellations, the majority of the stars in the Plough lie at the same distance and were born together as members of the Ursa Major Moving Group, which also includes **Menkalinan** in **Auriga** and Gemma in Corona Borealis. The exceptions are Dubhe and **Alkaid**, and – as these two stars go off on their own paths – the shape of the Plough will gradually change.

JANUARY'S OBJECT

Almost 143,000 kilometres in diameter, **Jupiter** is the biggest world in our Solar System and could contain 1300 Earths. It's made almost entirely of compressed gases, with liquid hydrogen near its centre

behaving like a metal and generating the strongest magnetic field of any planet.

Despite its vast size, Jupiter spins round in only 9 hours 55 minutes. As a result, its equator bulges outwards. Through a small telescope you can make out bright and dark bands stretched round the tangerine-shaped planet – white clouds of frozen ammonia overlying the planet's darker lower atmosphere – along with the Great Red Spot (see Picture).

Astronomers have found 95 moons orbiting Jupiter. The four biggest are visible in binoculars, and even – to the really sharp-sighted – with the unaided eye. Ganymede is larger than the planet Mercury, while Callisto has the oldest unchanged surface of any world. Io sports hundreds of active volcanoes. Europa's icy surface conceals a deep ocean which may be home to some kind of aquatic life.

If I asked you what colour the stars are, you'd naturally reply 'white'. But look more carefully at some of the glorious gems of winter, preferably with binoculars. You'll find **Betelgeuse** shines red, while **Capella** is yellow and **Rigel** sparkles in blue-white hues.

These colours are a cosmic thermometer. Amazingly, without knowing even a star's nature or its distance, you can take its temperature. The coolest stars are red, while orange and yellow stars are progressively hotter, with blue-white stars right at the top of the temperature scale.

For instance, Betelgeuse's ruddy hue tells us that its surface temperature is a 'mere' 3200°C, as compared to 5500°C for our yellow Sun. This coolish star glows only dully, like an expiring log fire, but its huge size – 760 times wider than the Sun – means its total luminosity is very high. That's why astronomers call it a 'red supergiant'.

Capella is hotter, so it has a yellow aura, like the Sun. And Rigel is near the top end of the stellar temperature scale, its blue-white surface at an incandescent 12,000°C. Once we plug in its distance, it turns out that this super-hot star shines 120,000 times more brilliantly than our Sun.

JANUARY'S PICTURE

Light and dark bands of clouds whirl around Jupiter's massive bulk, at different speeds depending on their latitude. And caught between them, like a ball bearing in a race, is the Great Red Spot. The largest storm in the Solar System, Jupiter's Great Red Spot rotates once every six days with winds at five times hurricane force.

Its distinctive hue, captured beautifully in this shot by Pete Lawrence, has long been a mystery. Astronomers now think it occurs because this storm rises well above the other cloud decks. Exposed to ultraviolet radiation from the Sun, simple gases in its upper levels are converted into more complex and colourful molecules. In other words, the Great Red Spot is a case of cosmic sunburn!

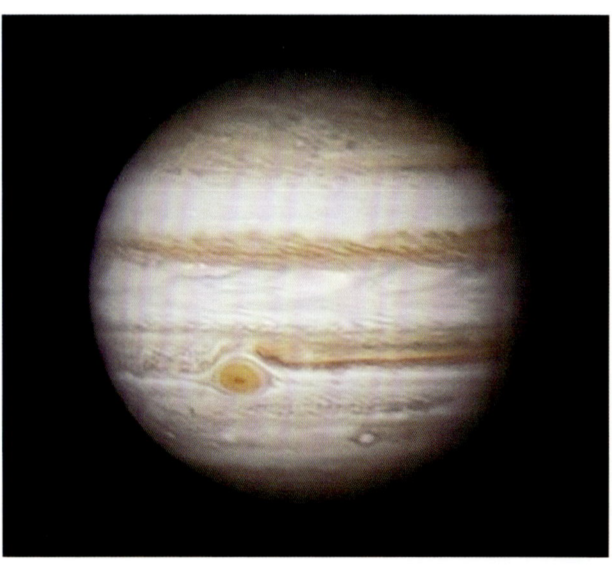

Observing from Thornton, Leicestershire, Pete Lawrence imaged Jupiter on 4 September 2023 using a Celestron C14 355-mm Schmidt-Cassegrain telescope at f/18 plus a Player One Uranus-C camera with a luminance filter. Pete stacked the sharpest frames from a 20-minute observing session, 'de-rotated' to account for the planet's rapid spin.

JANUARY'S CALENDAR

SUNDAY	MONDAY	TUESDAY	WEDNESDAY	THURSDAY	FRIDAY	SATURDAY
				1	2	3 10.03 am Full Moon near Jupiter; supermoon; Earth at perihelion; Quadrantids
4 Quadrantids (am); Moon near Praesepe	5 Moon very near Praesepe (am)	6 Moon near Regulus	7	8	9 Jupiter closest to Earth	10 3.48 pm Last Quarter Moon; Jupiter opposition
11 Moon near Spica (am)	12	13	14	15 Moon near Antares (am)	16	17
18 7.52 pm New Moon	19	20	21	22 Moon near Saturn	23 Moon near Saturn	24
25	26 4.47 pm First Quarter Moon	27 Moon occults the Pleiades	28	29	30 Moon near Jupiter	31 Moon near Jupiter (am)

SPECIAL EVENTS

- **3 January:** the first of this year's three supermoons (see December's Special Events and Topic) tracks just above Jupiter.
- **3 January, 5.15 pm:** the Earth is at perihelion, its closest point to the Sun (147 million km away).
- **Night of 3/4 January:** the maximum of the **Quadrantid meteor shower**, dust particles from the old comet 2003 EH_1 burning up in the Earth's atmosphere. Unfortunately, light from the Full Moon will wash out all but the brightest shooting stars.
- **Night of 4/5 January:** the Moon grazes the top of Praesepe.
- **9 January:** Jupiter is closest to the Earth at 633 million km.
- **10 January:** Jupiter lies opposite to the Sun in the sky.
- **22 January:** Saturn lies to the upper left of the crescent Moon.
- **23 January:** you'll find Saturn below the Moon.
- **27 January, 8–11.30 pm:** the Moon moves across the top of the Pleiades, hiding a couple of the brighter Seven Sister stars (Chart 1b).
- **Night of 30/31 January:** the Moon passes over Jupiter.

JANUARY'S PLANET WATCH

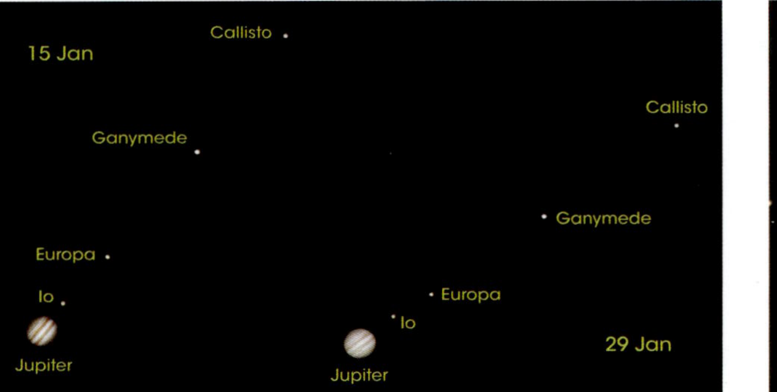

1a *15 January, 6 pm; 29 January 9 pm. Jupiter and its four largest moons in order (binocular view).*

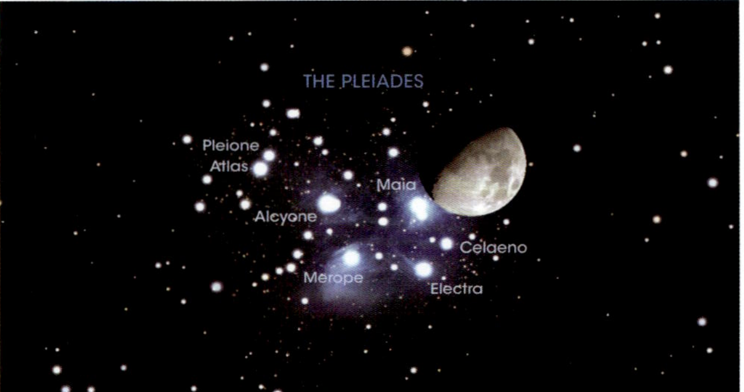

1b *27 January, 9.15 pm. The Moon occults the Pleiades.*

• Let's start with **Jupiter**, putting on its most magnificent display of the year as it passes closest to the Earth in January. Visible all night long in Gemini, the giant of the Solar System is shining more brilliantly than any of the stars, at magnitude –2.7, with a pure steady light. The Full Moon passes right above Jupiter on 3 January.

• Even a pair of binoculars will show you the planet's four largest moons, which are comparable in size to our Moon or the planet Mercury. Their positions change from night to night, but two good nights to identify them are 15 January (around 6 pm) and 29 January (about 9 pm), when the moons are arranged in order on the same side of Jupiter: Io, Europa, Ganymede and Callisto (Chart 1a).

• The only significant 'star' in the south-western evening sky is the planet **Saturn**, shining at magnitude +1.0 on the borders of Aquarius and Pisces, and setting around 10 pm. The crescent Moon is nearby on 22 and 23 January.

• **Neptune** is about three degrees above Saturn, in Pisces, and at magnitude +7.8 is visible only in binoculars or a telescope. It also sets about 10 pm.

• Its near twin, **Uranus**, is near the limit of naked-eye visibility at magnitude +5.7. The seventh planet lies in Taurus, below the Pleiades, and sinks below the horizon around 4 am.

• **Mercury, Venus** and **Mars** are too close to the Sun to be visible this month.

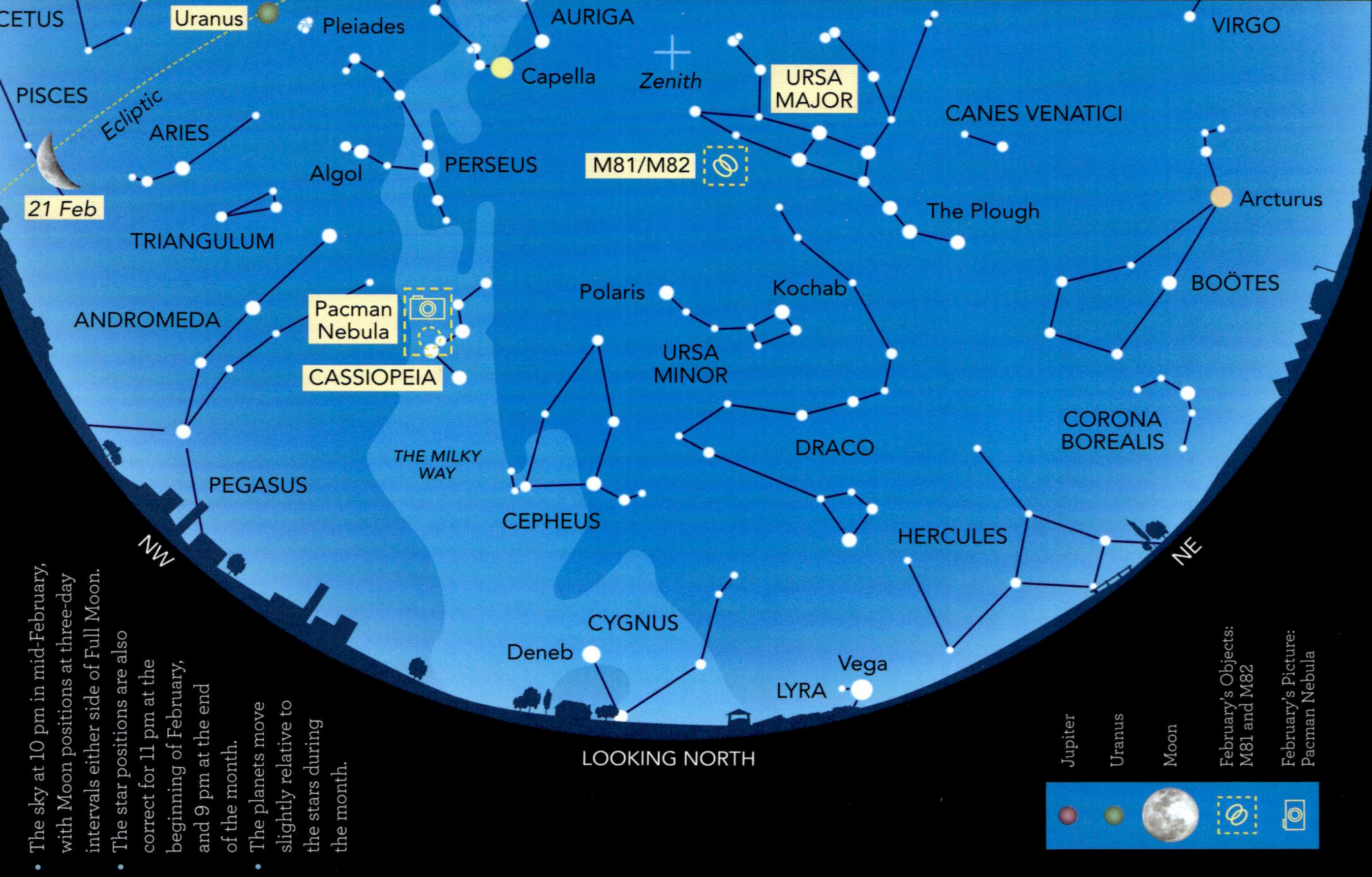

- The sky at 10 pm in mid-February, with Moon positions at three-day intervals either side of Full Moon.
- The star positions are also correct for 11 pm at the beginning of February, and 9 pm at the end of the month.
- The planets move slightly relative to the stars during the month.

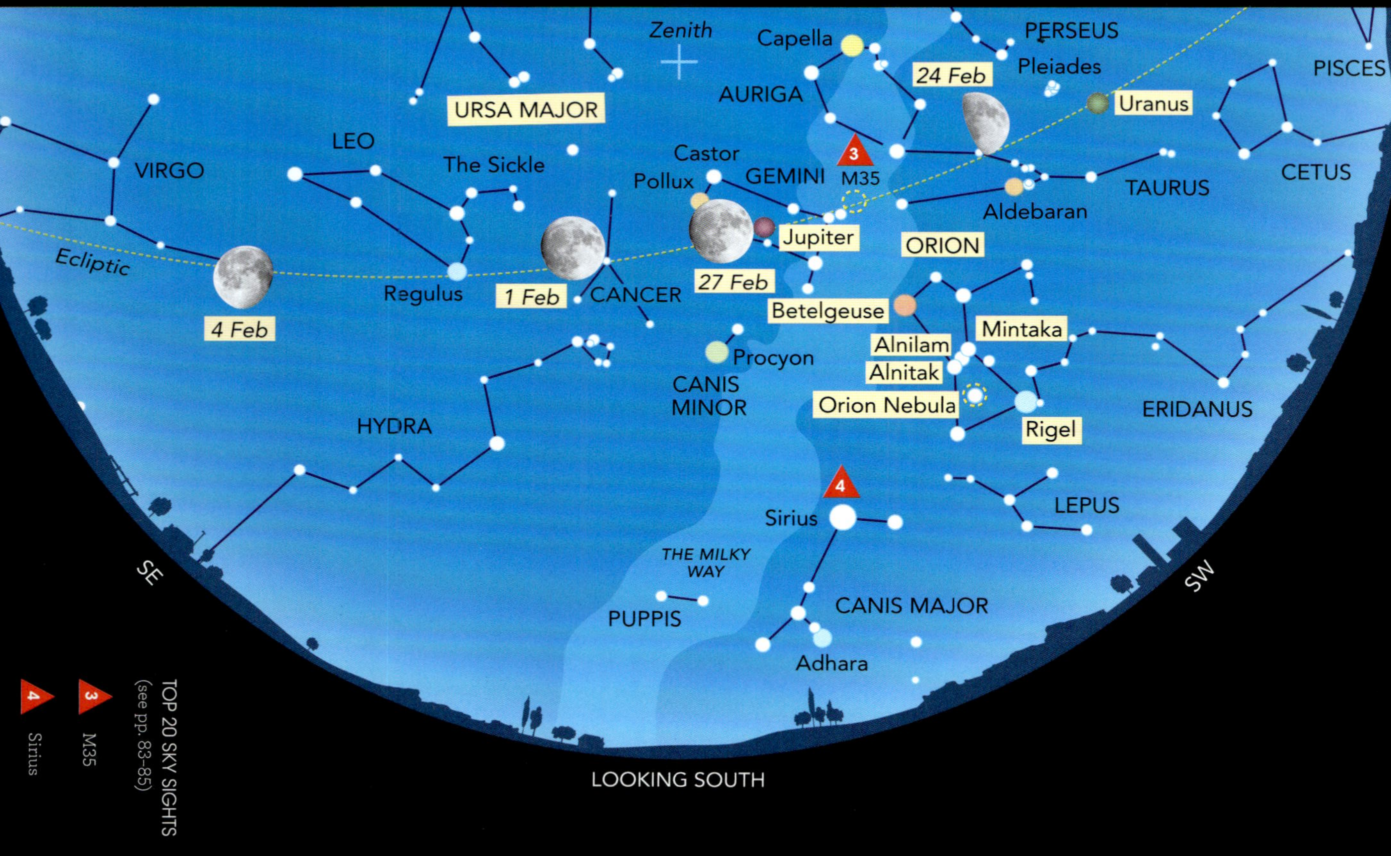

EAST
Zenith
Capella
PERSEUS
WEST
PISCES
Pleiades
24 Feb
Uranus
AURIGA
URSA MAJOR
CETUS
LEO
Castor
M35
3
TAURUS
VIRGO
The Sickle
Pollux
GEMINI
Aldebaran
Jupiter
ORION
Ecliptic
Regulus
1 Feb
CANCER
27 Feb
Betelgeuse
4 Feb
Mintaka
Alnilam
Alnitak
Procyon
Orion Nebula
CANIS MINOR
Rigel
ERIDANUS
HYDRA
4
LEPUS
Sirius
THE MILKY WAY
SE
SW
PUPPIS
CANIS MAJOR
Adhara
LOOKING SOUTH
4   Sirius
3   M35
(see pp. 83–85)
TOP 20 SKY SIGHTS
FEBRUARY
15
FEBRUARY

We have not just one Evening Star, but two! Venus is joined by elusive Mercury, putting on its best show of the year. Add in the four outer planets, and this month we have the chance to view all the worlds of the Solar System (bar Mars) in the evening sky.

FEBRUARY'S CONSTELLATION

The scintillating stellar figure of **Orion** strides high across the heavens this month. According to various ancient cultures, this constellation represented a shepherd, a mighty king or even a canoe (!); but to the ancient Greeks he was a formidable hunter, and also the world's most handsome man. In the sky Orion also cuts a dash, containing one-tenth of the brightest stars in the sky.

Blood-red **Betelgeuse** marks one of his shoulders. This red giant is 760 times wider than the Sun, and one day it's destined to blow itself apart as a supernova. Orion's other brilliant star, **Rigel**, lies at the bottom of the hunter's tunic. In contrast to Betelgeuse, it's blue-white in colour, with a searingly hot temperature of 12,000°C.

Orion's Belt is marked by well-matched **Alnitak**, **Alnilam** and **Mintaka**, lying over 1200 light years away. Despite this immense distance, they're prominent in our skies because each out-shines our Sun 200,000 times over.

Look carefully below the Belt to spot a faint patch of light representing Orion's sword. This is the great **Orion Nebula**, a seething maelstrom of incandescent gas that's the birthplace of new stars (see December's Object).

FEBRUARY'S OBJECT

A pair of galaxies this month, lying 12 million light years away in the constellation **Ursa Major**. M81 and M82 are visible through good binoculars on a really dark night, though you'll need a moderately powerful telescope to reveal them in detail.

M81 is a beautiful, smooth spiral galaxy with curving spiral arms wrapped around a softly glowing bulge of stars. At its heart lies a black hole 15 times heavier than the black hole at the centre of the Milky Way.

Though it's also a spiral galaxy, **M82** couldn't be more different. It looks a total mess, with a huge eruption taking place at its core. Some 300 million years ago, a close encounter with M81 ripped out streams of gas that are still raining back down onto M82's core, creating an explosion of star formation – making M82 the prototype 'starburst galaxy'.

FEBRUARY'S TOPIC: BIGGEST TELESCOPES

A backyard telescope with a lens or mirror 100 millimetres across is a good

---

OBSERVING TIP

It's best to view your favourite objects when they're well clear of the horizon. If you observe them low down, you're looking through a large thickness of the atmosphere – which is always shifting and turbulent. It's like trying to observe the outside world from the bottom of a swimming pool! This turbulence makes the stars appear to twinkle. Low-down planets also twinkle, although to a lesser extent, because they subtend tiny discs, and aren't so affected.

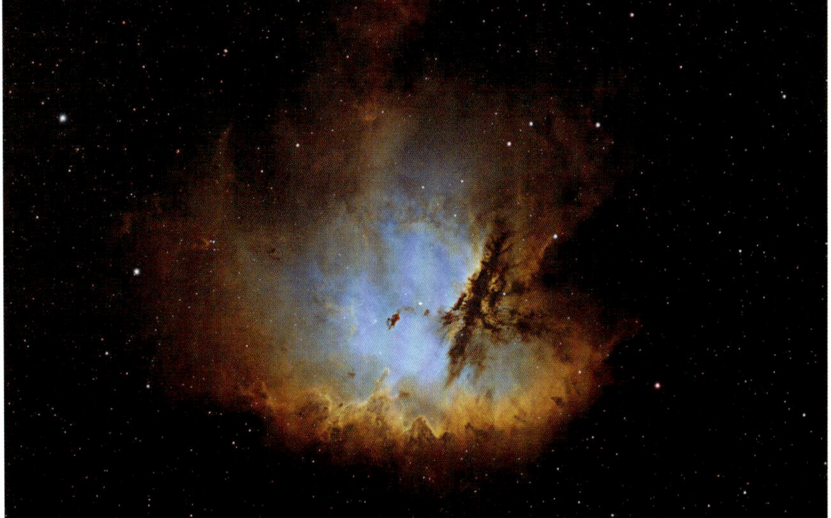

Peter Jenkins captured the Pacman Nebula from Nottinghamshire, using an Officina Stellare RH200 200-mm f/3 telescope and a ZWO ASI2600MM camera. He took 24 × 300-second exposures each through hydrogen alpha, oxygen light ([OIII]) and sulphur emission ([SII]) filters, for a total exposure of 6 hours. The image was processed in PixInsight and Photoshop.

instrument for general stargazing, but every astronomer has 'aperture fever' – the urge for the widest-possible instrument.

With a bigger aperture you can use a higher magnification, and also see fainter objects. However, refractors have a natural limit, as a really large lens will distort under its own weight. For over a century, the record-holder has been the 1-m Yerkes Great Refractor near Chicago, USA.

Reflectors have no such limit, as you can support a mirror from behind. The twin 8.4-m mirrors of the Large Binocular Telescope in Arizona, USA, currently hold the record. But astronomers have created even larger apertures out of a mosaic of small mirrors accurately fitted together to create a single reflecting surface. The biggest is the 10.4-m Gran Telescopio Canarias on La Palma in the Canary Islands.

But it will be overtaken in 2028 by the European Southern Observatory's latest instrument in Chile. Boasting a 39-m mirror made from 798 individual mirror tiles, it well deserves its name: the ELT, or Extremely Large Telescope.

FEBRUARY'S PICTURE

Defocus your eyes, and the glowing gas cloud NGC 281 looks a bit like the 1980s video-game character Pac-Man. But Peter Jenkins's superbly detailed image shows that there's far more to the **Pacman Nebula**.

Located 9000 light years away in **Cassiopeia**, this nebula and its central star (magnitude +7.8) are visible in a small telescope. Under higher power, you can see that the latter object is in reality a quintet of blue-white stars; far heavier than the Sun, their ultraviolet radiation is responsible for illuminating the whole nebula.

And Peter's picture clearly reveals that the Pac-Man 'mouth' to the right is the silhouette of a complex of dark clouds. The radiation from the massive central stars has compressed the interstellar material here to form a new generation of lower-mass stars, like the Sun.

SUNDAY	MONDAY	TUESDAY	WEDNESDAY	THURSDAY	FRIDAY	SATURDAY
1 10.09 pm Full Moon	**2** Moon near Regulus	**3**	**4**	**5**	**6** Moon near Spica	**7**
8	**9** 12.43 pm Last Quarter Moon	**10**	**11** Moon near Antares (am)	**12**	**13**	**14**
15	**16**	**17** 12.01 pm New Moon; annular solar eclipse	**18** Moon near Mercury and Venus	**19** Mercury E elongation; Moon near Saturn	**20** Saturn near Neptune	**21**
22	**23**	**24** 12.27 pm First Quarter Moon	**25**	**26** Moon near Jupiter	**27** Moon near Jupiter	**28** Moon near Praesepe

SPECIAL EVENTS

- **17 February:** an annular eclipse of the Sun crosses the coast of Antarctica; it's partial as seen from the rest of Antarctica, south-east Africa and the southern Indian Ocean, but nothing is visible from Britain and Ireland.
- **18 February, 6 pm:** look low in the twilight glow in the west to spot the lovely sight of the narrowest crescent Moon lying between Venus

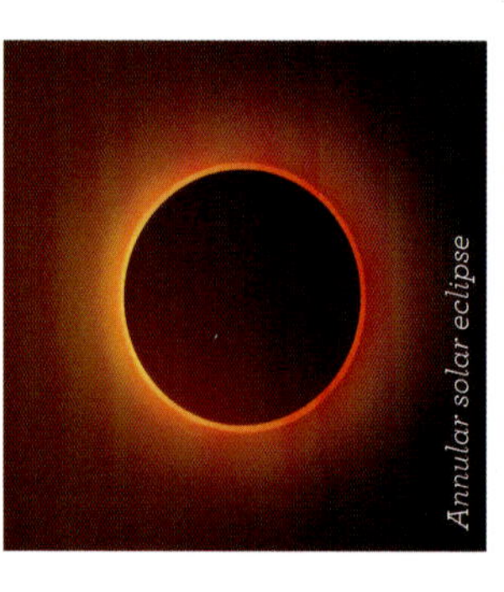

Annular solar eclipse

(below) and Mercury (above), with Saturn slightly higher in the sky (Chart 2a).

- **19 February:** Mercury is at its greatest separation from the Sun. You'll find it in the west after sunset; Venus lies below, while the crescent Moon teams up with Saturn above the two innermost planets (Chart 2a).
- **20 February:** Saturn passes close to Neptune (see Planet Watch and Chart 2b).
- **26 February:** Jupiter lies to the left of the Moon, with Castor and Pollux above.
- **27 February:** You'll find the Moon under Castor and Pollux, with the planet Jupiter to the right.
- **28 February:** Praesepe, also known as the Beehive Cluster, lies immediately below the Moon – a lovely sight in binoculars or a small telescope.

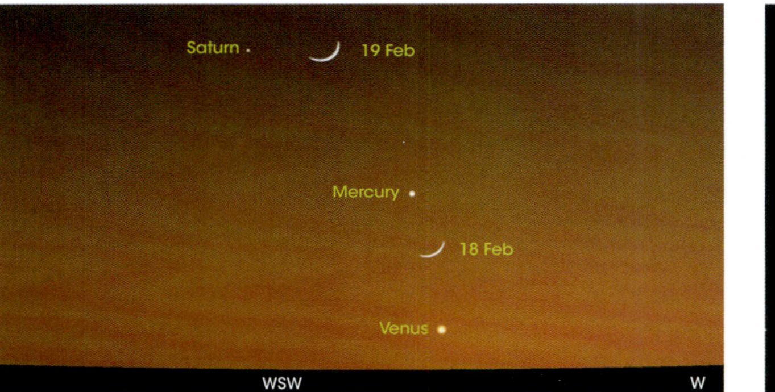

2a *18–19 February 6 pm. The crescent Moon passes Venus, Mercury and Saturn.*

2b *20 February, 7 pm. Saturn near Neptune.*

• At the beginning of February, **Saturn** is the only planet visible to the naked eye in the west after sunset. At magnitude +1.0, it lies in the dim constellation of Pisces and sets around 8 pm. The Moon is nearby on 19 February (Chart 2a).

• Also in Pisces, **Neptune** is 500 times fainter than Saturn at magnitude +7.8, and setting about 8 pm. Saturn starts the month below Neptune, but is gradually moving upwards and passes the outermost planet on 20 February. Around this time you have a great chance to identify the planet (in a telescope or good binoculars), as the faint greenish 'star' 50 arcminutes (almost two Moon-diameters) to the right of Saturn (Chart 2b).

• From about 8 February, you can spot **Mercury** well to the lower right of Saturn. The innermost planet is putting on its best evening performance of the year, gradually fading from magnitude –1.1 to –0.4 when Mercury reaches elongation on 19 February; it's then setting as late as 7 pm.

• Around the middle of February, **Venus** (magnitude –3.9) appears below Mercury, and on 18 and 19 February the Moon plays tag with Venus, Mercury and Saturn (Chart 2a). By the end of the month, Mercury has faded to magnitude +1,8, and Venus is dominating the evening sky, sinking below the horizon around 7.45 pm.

• **Uranus** lies in Taurus, at magnitude +5.8, and sets about 1.30 am.

• Giant planet **Jupiter** blazes at magnitude –2.5 in Gemini, setting around 5.30 am. The Moon is nearby on 26 and 27 February.

• **Mars** is lost in the Sun's glare this month.

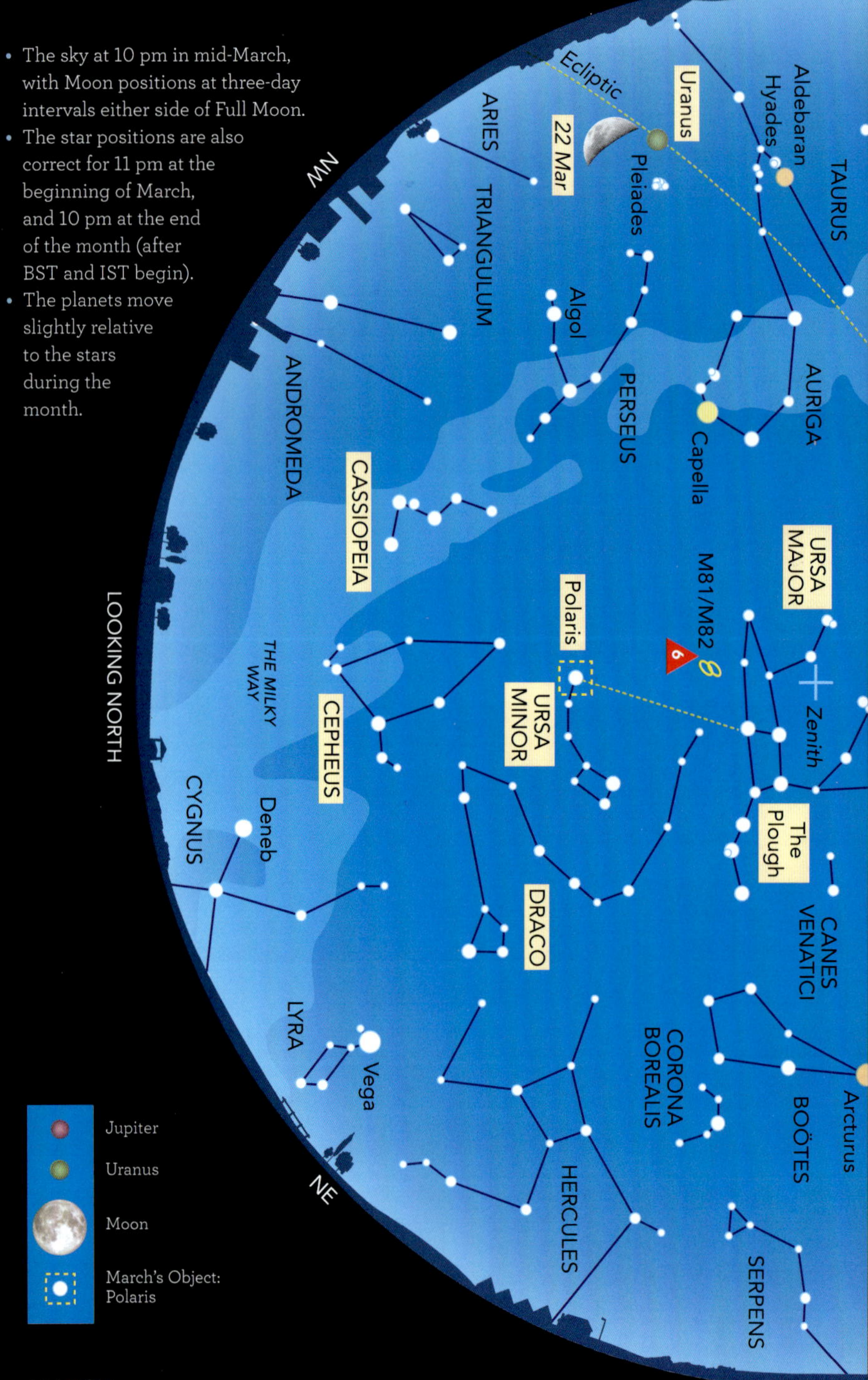

- The sky at 10 pm in mid-March, with Moon positions at three-day intervals either side of Full Moon.
- The star positions are also correct for 11 pm at the beginning of March, and 10 pm at the end of the month (after BST and IST begin).
- The planets move slightly relative to the stars during the month.

MARCH

EAST
WEST
SE
SW
LOOKING SOUTH
BOÖTES
SERPENS
Arcturus
CANES VENATICI
Zenith
URSA MAJOR
AURIGA
25 Mar
M35
Aldebaran
TAURUS
Castor
Pollux
Jupiter
GEMINI
The Sickle
LEO
Praesepe
5
Betelgeuse
ERIDANUS
VIRGO
CANCER
ORION
Rigel
28 Mar
Regulus
Procyon
CANIS MINOR
Ecliptic
3 Mar
THE MILKY WAY
LEPUS
Spica
6 Mar
Sirius
CORVUS
HYDRA
CANIS MAJOR
PUPPIS
TOP 20 SKY SIGHTS
(see pp. 83–85)
5   Praesepe
6   M81 and M82

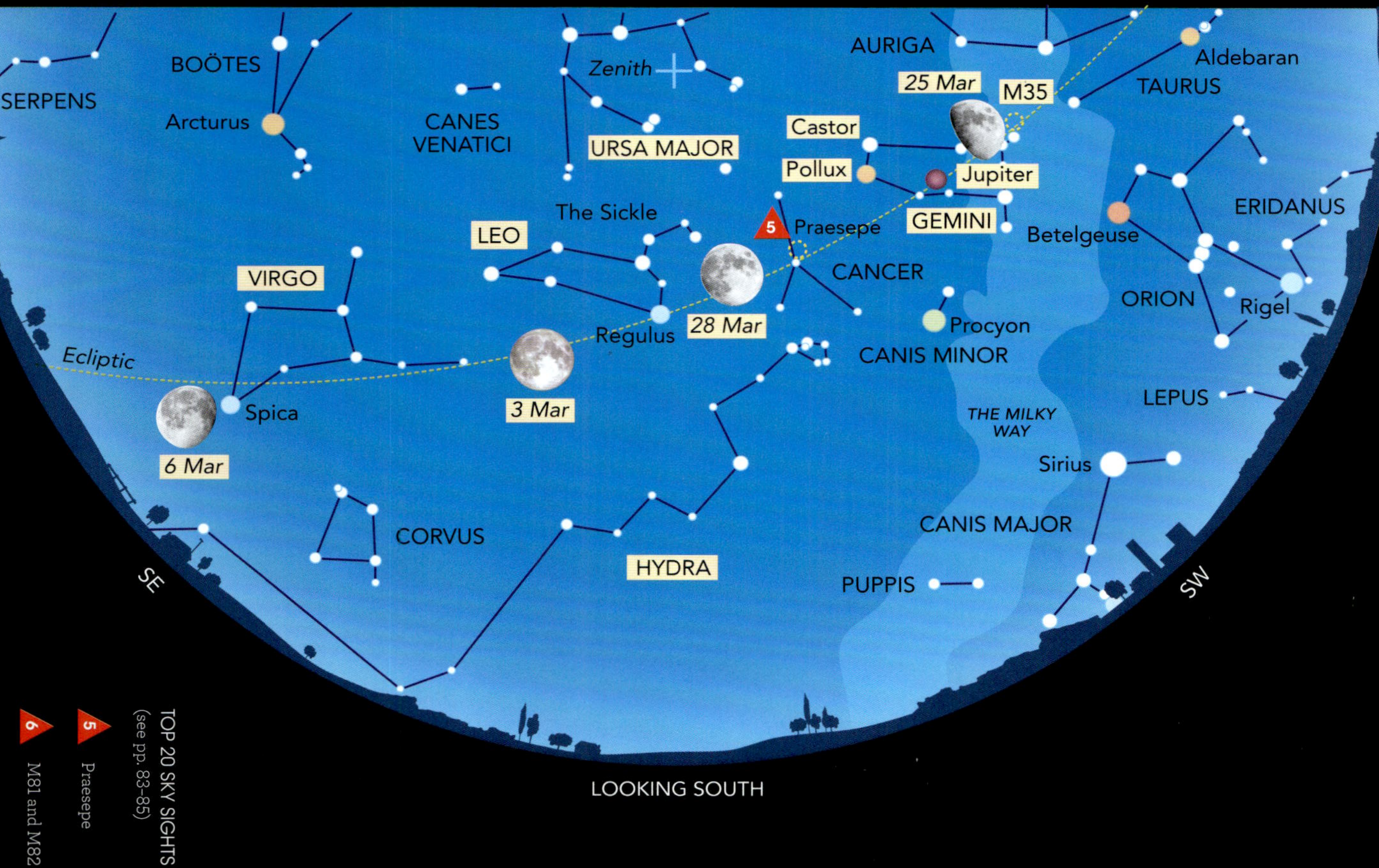

The two most brilliant planets, Venus and **Jupiter**, dominate these March evenings, joined by the large spring constellations **Leo**, **Virgo** and **Hydra** in the southern sky. And in the north we have the year-round constellations: Queen **Cassiopeia** and her consort **Cepheus**, **Draco** (the Dragon), and the two Bears – **Ursa Major** and **Ursa Minor.**

MARCH'S CONSTELLATION

The constellation of the celestial Twins – **Gemini** – is crowned by the bright stars Castor and Pollux, representing the heads of the celestial brethren. In legend they were conceived by princess Leda on the same day, Castor by her human husband, and immortal Pollux by chief god Zeus. After Castor's death, Zeus placed them together for eternity among the stars.

Castor is an amazing family of six stars. Through a small telescope, you can see that it's a close double, with a fainter companion lying further away. And all three of these stars are themselves very tight-knit pairs. **Pollux** – cooler and orange in hue – is the nearest giant star to the Sun. Gemini also boasts a pretty star cluster, **M35**.

MARCH'S OBJECT

Many people expect the famous Pole Star, or North Star, to be one of the brightest stars. But in fact, **Polaris** (to use its official name) is surprisingly modest: at magnitude +2.0 it's similar to the stars making up the **Plough**, and you can use the end stars of the Plough to locate it (see Star Chart).

Polaris just happens to lie above the Earth's North Pole, so our planet rotates underneath it. As a result, this star remains almost stationary in the sky, due north wherever or whenever you observe it. That's why William Shakespeare has

Julius Caesar declare in his eponymous play: 'I am constant as the Northern Star, Of whose true fixed and resting quality There is no fellow in the firmament.'

Polaris lies at the end of the tail of the Little Bear (**Ursa Minor**). It's a massive yellow star, five times heavier than the Sun and 1200 times brighter. As it swells and shrinks regularly, the North Star's brightness varies slightly over a period of four days.

MARCH'S TOPIC: GALILEO GALILEI

The Italian scientist Galileo (1564–1642) was one of the all-time greats of astronomical history, but – amazingly – most of his celestial breakthroughs occurred in just one miraculous year, from autumn 1609 to autumn 1610.

It began when Galileo heard of an 'optick tube' invented in the Netherlands that made distant objects appear nearer. A skilled craftsman himself, Galileo constructed a much better telescope with his own hands. In the space of a few weeks, he had observed craters on the Moon, innumerable stars making up the Milky Way and the four largest moons of Jupiter. Other astronomers saw all these sights around the same time, but Galileo was the first to rush his observations into print.

His motives were twofold: instant fame and riches; and to prove that the Earth orbits the Sun. Critics had argued that – if the Earth moved – it would leave the Moon behind. Now, Galileo could say that Jupiter carries its moons with it, so undoubtedly the Earth could as well. A little later, he discovered that Venus shows phases like the Moon, and that could only happen if Venus orbits the Sun rather than the Earth.

The rest of his life was devoted to experiments in physics. Albert Einstein lauded Galileo as 'the father of modern physics – indeed, of modern science altogether'.

MARCH'S PICTURE

The Sun's magnetic weather has sculpted the details in this close-up image of a 'solar active region' taken by Alexandra Hart on 16 February 2014, when our local star was near the maximum of its 11-year cycle of activity.

Magnetic lines of force are holding up the prominences – curtain-like bands of glowing gas – which tower upwards for several times the Earth's diameter. At the lower left, intense magnetic fields are inhibiting the energy that bubbles up from the Sun's core, to create a clutch of dark sunspots. They are rimmed by brighter regions – 'plages' – where the suppressed energy struggles out.

And the magnetism shapes the gases around each sunspot into dark and bright whorls, like the patterns you see when you scatter iron filings around a magnet: Alexandra calls these 'ripples in a pond' – though on a scale where a ripple would envelope our entire planet!

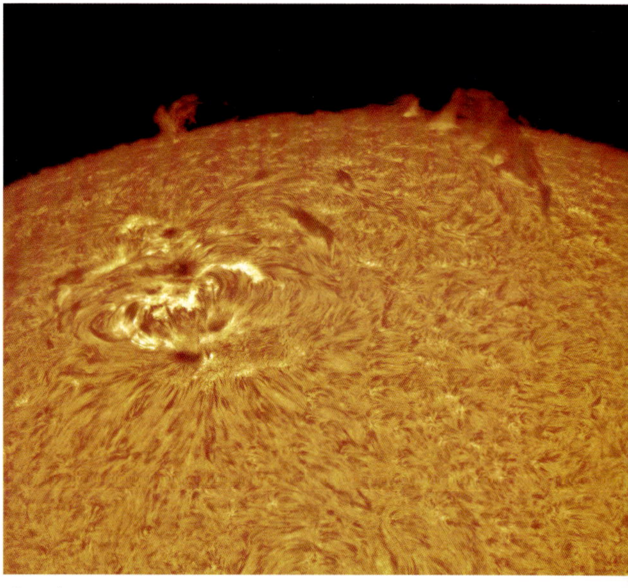

The key to Alexandra Hart's superb image of solar active region 11974 was a Solarscope DS 100-mm filter which passes just the red light emitted by hydrogen alpha. She attached it to a TEC 140 140-mm refractor equipped with a 2× Barlow lens and a PGR Grasshopper3 camera. The best 10 per cent of around 1000 images were stacked using AutoStakkert! 2. The image was sharpened using Photoshop CS5, then false colour added.

SUNDAY	MONDAY	TUESDAY	WEDNESDAY	THURSDAY	FRIDAY	SATURDAY
1	2 Moon near Regulus	3 11.38 am Full Moon; total lunar eclipse	4	5	6 Moon near Spica	7 Venus near Neptune
8 Venus near Saturn	9	10 Moon near Antares (am)	11 9.38 am Last Quarter Moon	12	13	14
15	16	17	18	19 1.23 am New Moon	20 Spring Equinox; Moon near Venus	21
22	23	24	25 7.18 pm First Quarter Moon	26 Moon near Jupiter	27	28
29 BST and IST begin (am); Regulus occultation	30	31				

SPECIAL EVENTS

- **3 March:** people in the Americas, Asia and Australasia are treated to a total eclipse of the Moon, but nothing is visible from Africa or Europe, including Ireland and the UK.
- **7 March:** Venus passes close to Neptune – outshining it 50,000 times over – with Saturn nearby (see Planet Watch).
- **8 March:** Saturn is only 55 arcminutes to the left of Venus.
- **20 March, 2.46 pm:** the Spring Equinox, when day and night are equal.
- **20 March:** the crescent Moon and Venus form a stunning duo low in the west after sunset (Chart 3a).
- **26 March:** you'll find Jupiter below the Moon, with Castor and Pollux above.
- **29 March, 1.00 am:** British Summer Time and Irish Standard Time start – put your clocks forward tonight.
- **29 March, 8.20 pm:** Regulus reappears from behind the Moon (the exact time depends on your location). Through binoculars, you will see the star's companion, Regulus B (magnitude +8.1), emerge from occultation first. It's followed 6 minutes later by Regulus itself, 500 times brighter at magnitude +1.4 (Chart 3b).

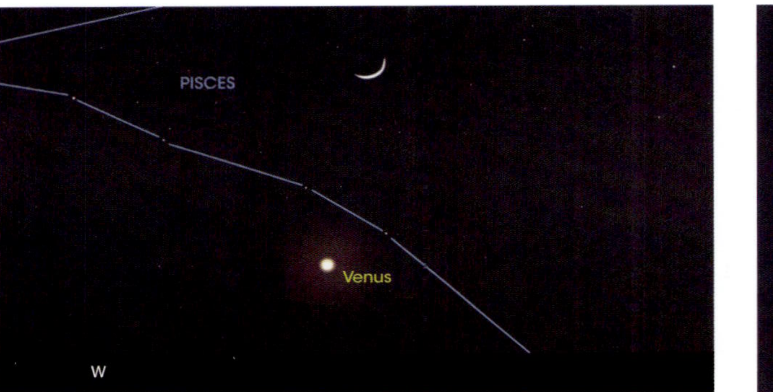

3a 20 March, 7.30 pm. Venus with the crescent Moon.

3b 29 March, 8.20 pm. Regulus emerges from occultation.

- **Venus** is a brilliant Evening Star, shining at magnitude –3.9 low in the west after sunset like a lantern in the dusk twilight. It sets around 7.30 pm. The crescent Moon forms a stunning sight with Venus on 20 March (Chart 3a).
- On the first couple of evenings of March, you may catch **Mercury** (magnitude +2.4) to the right of Venus and setting about 7 pm, but it quickly sinks into the twilight glow and disappears.
- At the beginning of the month, **Saturn** (in Pisces) lies above Venus, a hundred times fainter at magnitude +1.0, and setting around 7.30 pm. As the Evening Star rises in the sky, it passes less than a degree to the right of Saturn on 8 March. By mid-month, Saturn – like Mercury – has dropped down and out of sight into the Sun's glare.
- **Neptune** is just to the right of Saturn, also in Pisces and setting about 7.30 pm. At magnitude +7.8 you'll need a telescope to view it. As Venus zooms upwards in the sky, it passes only 4 arcminutes from Neptune on 7 March at 11.30 am, but this unusually close conjunction of the nearest and furthest planets is visible only from eastern Asia: they have separated to 20 arcminutes by the time the conjunction can be seen from the UK and Ireland.
- **Uranus** (magnitude +5.8) lies in Taurus below the Pleiades and sets about midnight.
- Still lording it over the night-time sky, **Jupiter** blazes in Gemini at magnitude –2.3, setting around 4 am. The Moon is just above Jupiter on 26 March.
- **Mars** is too close to the Sun to be visible this month.

- The sky at 11 pm in mid-April, with Moon positions at three-day intervals either side of Full Moon.
- The star positions are also correct for midnight at the beginning of April, and 10 pm at the end of the month.
- The planets move slightly relative to the stars during the month.

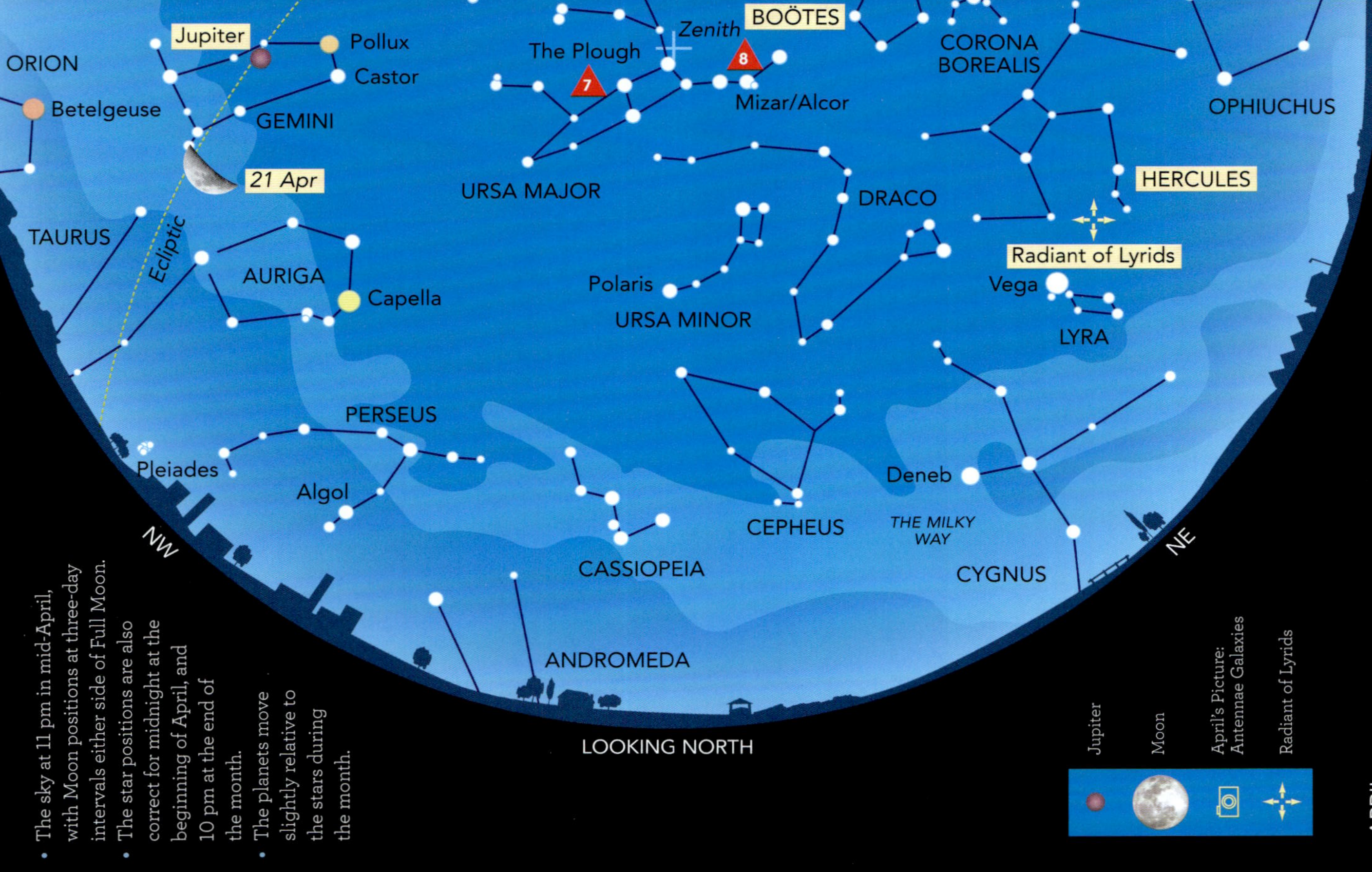

APRIL

EAST
WEST
8
Mizar/Alcor
7
The Plough
Castor
Pollux
HERCULES
CORONA BOREALIS
Zenith
GEMINI
THE MILKY WAY
OPHIUCHUS
CANES VENATICI
URSA MAJOR
CANCER
BOÖTES
Procyon
Arcturus
The Sickle
CANIS MINOR
Denebola
Algieba
24 Apr
SERPENS
LEO
Regulus
VIRGO
Spica
27 Apr
Alphard
Ecliptic
CORVUS
2 Apr
Antennae Galaxies
HYDRA
LIBRA
SE
SW
LOOKING SOUTH
TOP 20 SKY SIGHTS
(see pp. 83–85)
8 Mizar and Alcor
7 The Plough
APRIL
27

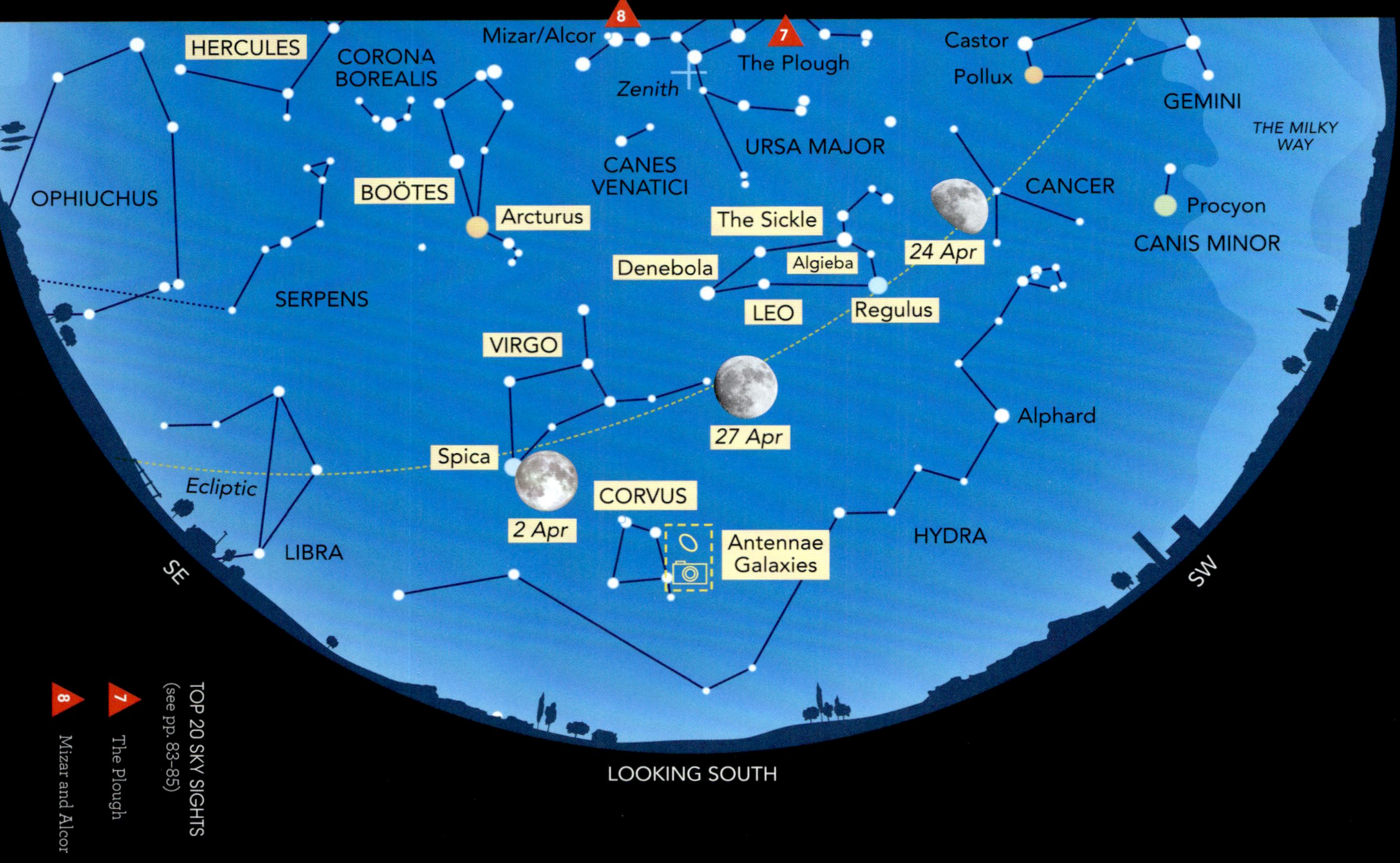

Venus and **Jupiter** adorn the winter constellations that are now setting in the west. In the east, three bright stars ride high this month, the principal players in the spring constellations. Leading the way is **Regulus** in **Leo**, with **Virgo**'s principal star **Spica** to the lower left, and orange **Arcturus** in **Boötes** lying above.

APRIL'S CONSTELLATION

Leo is one of the rare constellations that resembles the real thing – in this case, a crouching lion – and is one of the oldest constellations. In Greek mythology, Leo commemorated a fearsome beast that **Hercules** slaughtered as the first of his 12 labours, but these stars were first associated with a lion two millennia earlier, by the Elam civilisation on the shores of the Persian Gulf.

The lion's heart is marked by the first magnitude star **Regulus**. This celestial whirling dervish spins around in just 16 hours, making its equator bulge remarkably. Rising upwards is the **Sickle**, a back-to-front question mark that delineates the front quarters, neck and head of Leo. A small telescope shows that **Algieba**, the star that makes up the lion's shoulder, is a beautiful close double star.

The other extremity of Leo is marked by **Denebola**, meaning 'the lion's tail' in Arabic. Just underneath the feline's tummy is a clutch of spiral galaxies; too faint to be seen with the unaided eye, you can track them down with a small telescope.

APRIL'S OBJECT

Venus – the planet of love – is resplendent in our evening skies this month. Her pure lantern-like luminosity is beguiling. But looks are deceptive. Though Earth's twin in size, Venus could hardly be more different from our warm, wet world.

Venus is entirely draped in highly reflective clouds, but – unlike our friendly water clouds – Venus's all-enveloping pall is made of drops of concentrated sulphuric acid. They hang in an atmosphere of unbreathable carbon dioxide, so dense that the atmospheric pressure at its surface is 90 Earth-atmospheres. This thick blanket of carbon dioxide traps the Sun's heat, creating a runaway Greenhouse Effect that has made Venus the hottest planet in the Solar System: at 460°C, it's far hotter than an oven.

The surface bristles with 1600 major volcanoes, and maybe as many as a million smaller volcanic vents. Any way you look at it, seductive Venus is the planet from Hell!

APRIL'S TOPIC: ZODIACAL LIGHT

At this time of year, the Zodiac rises steeply from the horizon at dusk, making it an ideal month to catch one of the most

OBSERVING TIP

It's always fun to search out 'faint fuzzies' in the sky – star clusters, nebulae and galaxies. But don't even think of observing these objects around the time of Full Moon, as its light will drown them out. You'll have the best views near New Moon: a period astronomers call 'dark of Moon'. When the Moon is bright, focus on planets, prominent double stars – and, of course, the Moon itself.

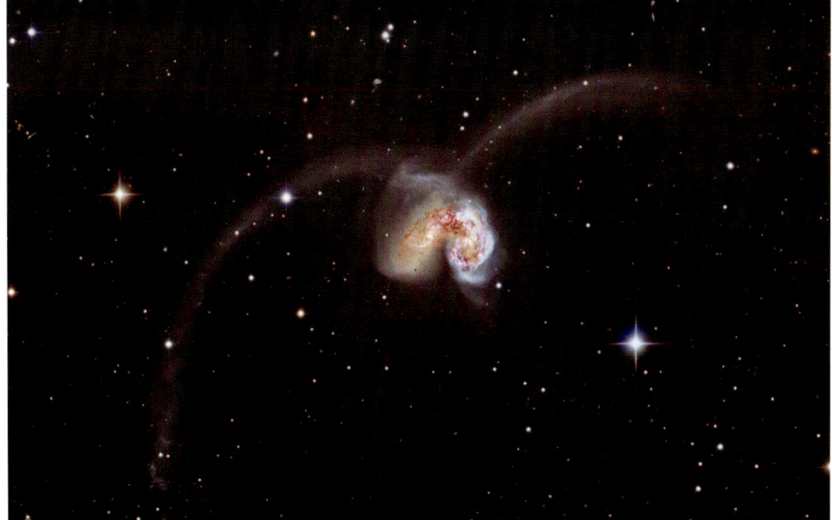

elusive of astronomical sights. Get away from streetlights, on a moonless night, and look to the west to try and spot a faint pyramid of light just after the Sun has gone down. This ghostly glow, following the line of the ecliptic (the path of the Sun, Moon and planets), is the zodiacal light.

When it's seen in the morning sky, the zodiacal light is often called the 'false dawn'. Its appearance over the Persian desert was vividly described by the 12th-century poet and astronomer Omar Khayyam: 'when false dawn streaks the east with cold, grey line, Pour in your cups the pure blood of the vine'.

The zodiacal light is caused by sunlight reflecting from a fog of tiny particles filling the inner Solar System. Most astronomers believe these dust particles are the remains of old comets and asteroids that have broken up, but there's recent evidence they may instead come from the dusty planet Mars.

APRIL'S PICTURE

Some 45 million light years away in the constellation **Corvus**, a slow-motion collision is unfolding before our eyes. The paths of two stately spiral galaxies

Neil Corke captured the Antennae Galaxies remotely, using a PlaneWave CDK24 f/6.5 610-mm catadioptric telescope at the Observatorio El Sauce in Chile. The camera was a Moravian Instruments C3-61000 PRO; he took 25 × 900-second exposures for luminance, plus exposures through four filters – 24 × 900-second (blue), 17 × 900-second (green), 17 × 900-second (red) and 15 × 900-second (hydrogen alpha) – for a total exposure of 24 hours 30 minutes.

have crossed in space, and the wreckage – stars and gas clouds – has spilt out into long streamers that have given this co-mingling pair their name: the **Antennae Galaxies**.

Neil Corke's crystal-clear photo reveals mayhem at the impact site. Clouds of gas compressed by the collision have broken out in a rash of starbirth, creating nebulae that shine brilliantly in the red light of hydrogen. Eventually, this merger will settle down to become an elliptical galaxy, a giant ball of stars largely devoid of gas and dust.

This image provides a preview of the fate of our own Galaxy. Five billion years in the future, the Milky Way will smash into the Andromeda Galaxy and the two will coalesce into a giant elliptical galaxy nicknamed Milkomeda.

APRIL'S CALENDAR

SUNDAY	MONDAY	TUESDAY	WEDNESDAY	THURSDAY	FRIDAY	SATURDAY
			1	2 — 3.12 am Full Moon near Spica	3	4 — Mercury W elongation
5	6	7 — Moon near Antares (am)	8	9	10 — 5.51 am Last Quarter Moon	11
12	13	14	15	16	17 — 12.52 pm New Moon	18 — Moon near Venus
19 — Moon near Venus and the Pleiades	20	21	22 — Moon near Jupiter; Lyrids	23 — Lyrids (am); Venus near Uranus	24 — 3.32 am First Quarter Moon	25 — Moon very near Regulus
26	27	28	29	30		

SPECIAL EVENTS

- **18 April:** the narrowest crescent Moon lies to the lower right of Venus, with Aldebaran and the Pleiades above (Chart 4a).
- **19 April:** Venus and the Moon team up in a beautiful tableau in the western sky as the sky grows dark, with Aldebaran to the left.

Look close to the Moon – preferably with binoculars – and you'll find the Pleiades right below the lunar crescent (Chart 4a).

- **22 April:** the bright 'star' to the left of the Moon is giant planet Jupiter, with Castor and Pollux above.
- **Night of 22/23 April:** shooting stars from the **Lyrid** meteor shower will be at their best in the early morning sky, after the Moon has set at 2 am. These meteors – debris from Comet Thatcher – often leave dusty trails as they burn up in the Earth's atmosphere.
- **23 April:** Venus passes Uranus to its lower left with the Pleiades to the upper right (Chart 4b).

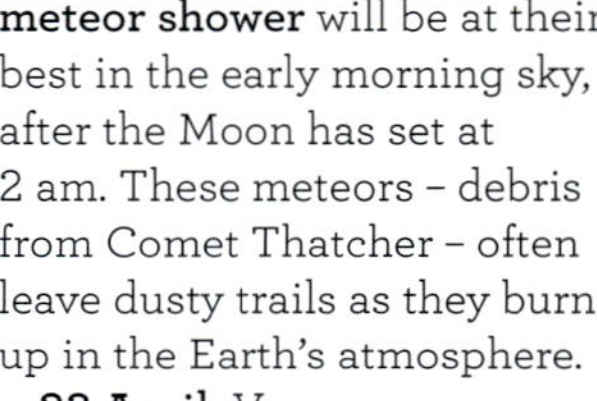

Lyrids

4a *18–19 April, 9.30 pm. The Moon passes Venus and the Pleiades.*

4b *23 April, 10 pm. Venus near Uranus.*

- This month, the glorious Evening Star rises high enough that you can see it against a truly dark sky: by the end of April, **Venus** is setting as late as 11 pm. At magnitude –3.9, it's brighter than anything else in the night sky bar the Moon.
- The crescent Moon lies near Venus on 18 and 19 April (Chart 4a). From 22 to 24 April, Venus passes to the left of the Pleiades.
- On 23 April, Venus lies to the right of **Uranus**, providing a convenient signpost for locating the seventh planet, which is on the borderline of naked-eye visibility at magnitude +5.8: in Taurus, it sets around 11 pm. On that date, use binoculars or a low-power telescope to find Uranus as a faint bluish-green 'star' 50 arcminutes (slightly less than two Moon-diameters) to the left of Venus and 8000 times fainter (Chart 4b).
- **Jupiter** makes its home in the stars of Gemini this

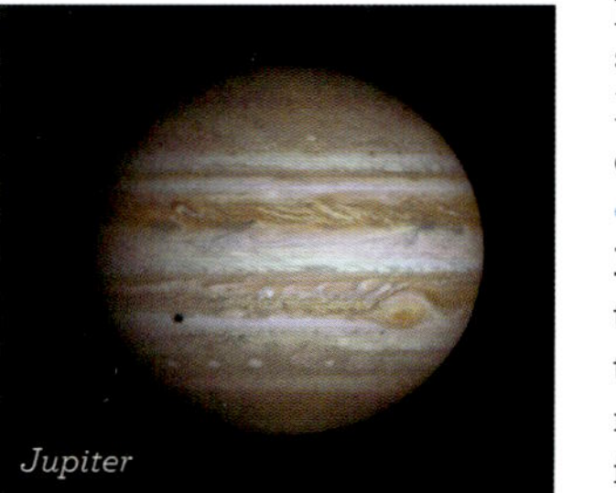

Jupiter

month. At magnitude –2.3, the giant planet is second in brightness only to Venus in the planetary canon, and more prominent than any of the stars. It sets about 3 am. The Moon is to the right of Jupiter on 22 April.
- **Saturn**, **Neptune**, **Mars** and **Mercury** are lost in bright twilight this month, even though the innermost planet reaches its greatest separation from the Sun on 4 April.

APRIL'S PLANET WATCH

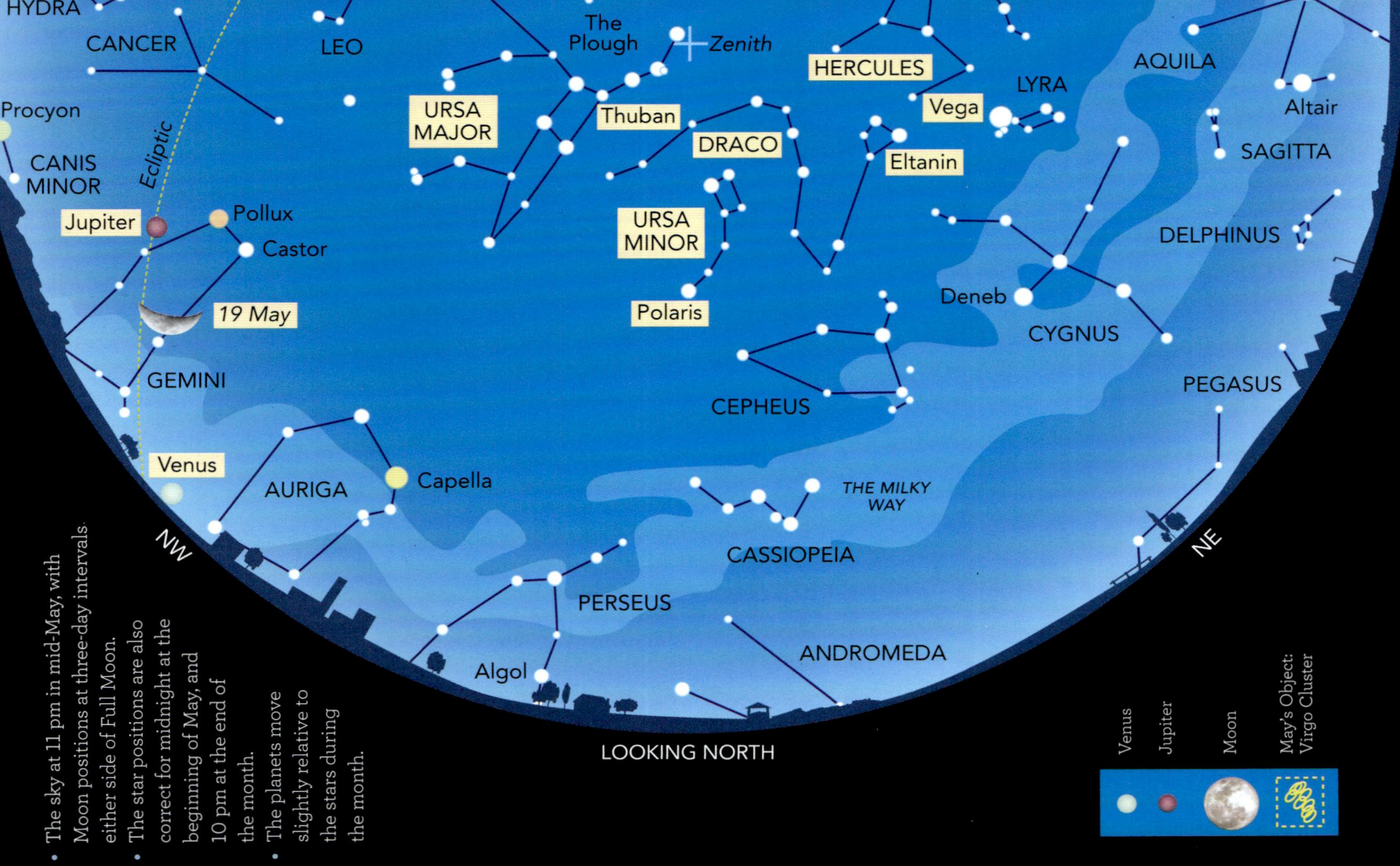

- The sky at 11 pm in mid-May, with Moon positions at three-day intervals either side of Full Moon.
- The star positions are also correct for midnight at the beginning of May, and 10 pm at the end of the month.
- The planets move slightly relative to the stars during the month.

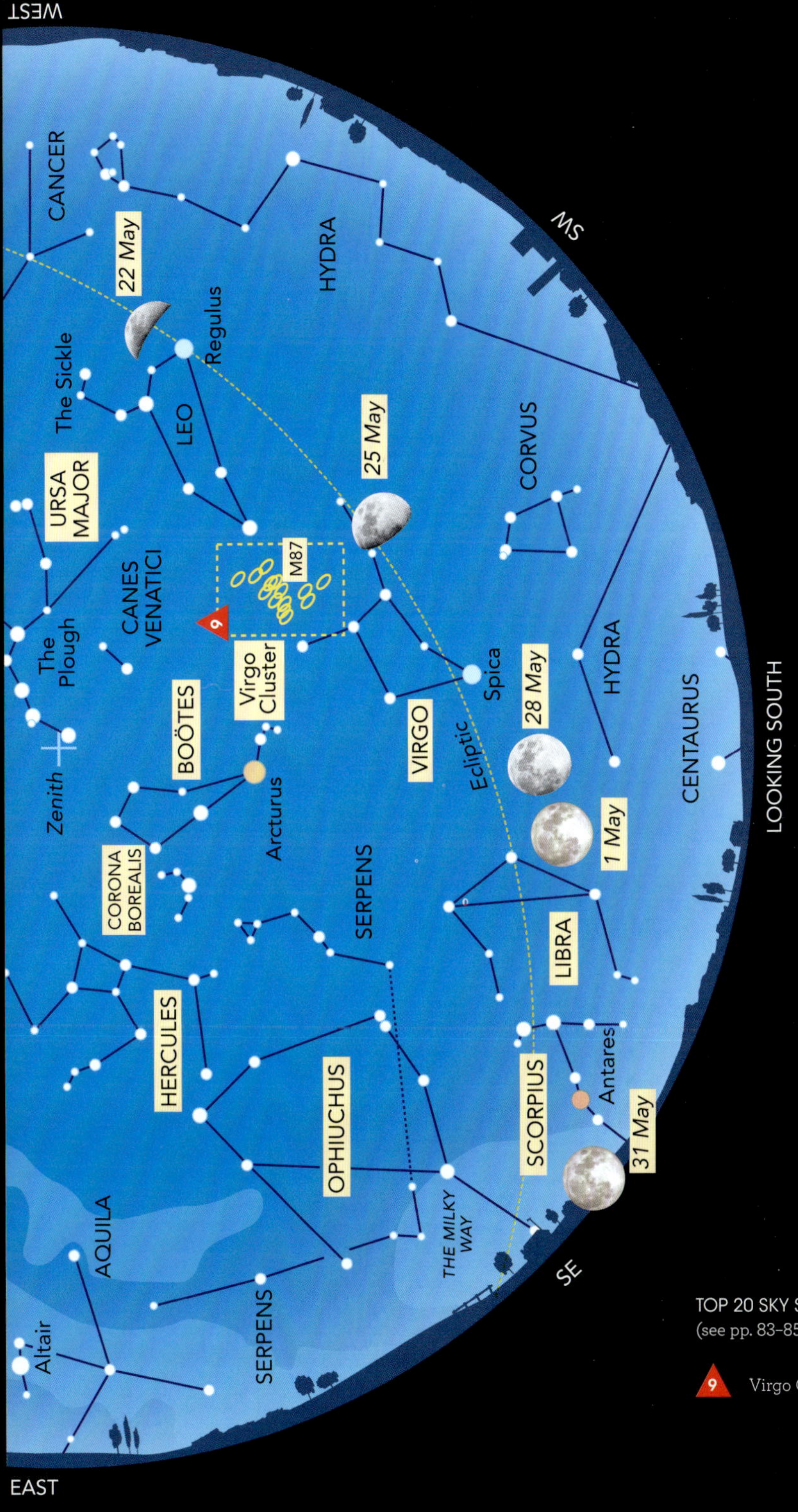

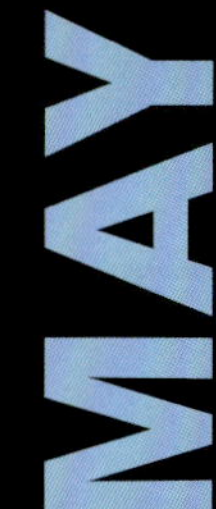

TOP 20 SKY SIGHTS
(see pp. 83–85)

9 Virgo Cluster

Tear your eyes away from **Venus** and **Jupiter**, twin planetary beacons in the May sky, to make the acquaintance of the early summer constellations: **Boötes**, with the little circlet of **Corona Borealis** to its left; **Scorpius**, **Libra** and **Virgo** to the south; and two faint sprawling giants – **Hercules** and **Ophiuchus** – in the south-eastern sky.

MAY'S CONSTELLATION

Draco, the cosmic dragon, writhes between the two bears (**Ursa Major** and **Ursa Minor**) in the northern sky. In Greek myth, Draco was slain by **Hercules** in his mission to steal golden apples from a sacred grove.

Eltanin, in the dragon's head, is an orange star 154 light years away, shining at magnitude +2.2. But in 1.5 million years' time, it will dash past the Earth at just 28 light years, blazing as the brightest star in our skies.

At first sight, **Thuban** seems an unassuming star, at mere magnitude +3.7. But what Thuban lacks in brightness, it makes up for in fame. Thuban was the Pole Star at the time the Great Pyramids were built in Egypt, in the third millennium BC. That's because the Earth's axis gradually swings around in space over a period of 26,000 years. At the moment, our planet's axis points to **Polaris** (see March's Object), but in 2800 BC it was lined up with Thuban. Come AD 13,700, brilliant **Vega** takes over as our North Star.

MAY'S OBJECT

If you have a small telescope, sweep the 'bowl' formed by **Virgo**'s Y-shape, and you'll detect dozens of fuzzy blobs. These are just the brightest members of the **Virgo Cluster**, our closest giant galaxy cluster, some 54 million light years away.

Large telescopes reveal 2000 galaxies in this vast swarm of star cities. Many of them are spirals like our Milky Way, while others are massive elliptical galaxies. The king of the cluster is **M87**, a giant elliptical with a massive central black hole that is ejecting a 5000-light-year-long jet of high-speed subatomic particles.

The Virgo Cluster is so massive that its gravity holds many neighbouring galaxy clusters in thrall, making it the centre of the Virgo Supercluster that's over 110 million light years across. The Milky Way – along with the rest of our Local Group of galaxies – lies in the outer suburbs of the Virgo Supercluster.

MAY'S TOPIC: BLUE MOON

On 31 May, the media headlines will be trumpeting the appearance of a 'blue Moon' in our skies. But don't expect an azure hue to our cosmic companion: the

OBSERVING TIP

Venus is a real treat this month, brilliant to the naked eye and – if you get to view it through a telescope – appearing as a small globe three-quarters lit up by the Sun. But don't wait for the sky to get totally dark, because the cloud-wreathed world will be too bright to make out any details. You're best off viewing Venus when the Evening Star first becomes visible in the twilight glow. Through a telescope, the planet then appears less dazzling against a pale blue sky.

lunar orb will appear just as white as any other Full Moon. It derives from the phrase 'once in a blue Moon', as only every two or three years do we witness two Full Moons in the same month.

The 'blue' tag wasn't devised by astronomers but appeared in an agricultural annual publication, the *Maine Farmers' Almanac*, in 1937. When there were four Full Moons in a three-month season, this almanac listed the third of them as a 'blue Moon'.

But a writer in the prestigious astronomy magazine *Sky & Telescope* misinterpreted the almanac and described the second Full Moon in a calendar month as a blue Moon. This usage caught on universally after 1986, when it became an answer in the popular quiz game Trivial Pursuit.

MAY'S PICTURE

The autumn skies of 2024 were enlivened by a rare celestial visitor: Comet Tsuchinshan–ATLAS hanging like a spectral dagger in the sky after sunset. The brightest comet widely visible from the northern hemisphere since Hale–Bopp in 1997 (Comet McNaught was amazing in southern skies in 2007), Comet Tsuchinshan–ATLAS was discovered early in 2023 by observatories in China and South Africa. It was falling towards the Sun from the homeland of comets, the Oort Cloud of giant icebergs that surrounds the Solar System.

Approaching the Sun, the comet was subjected to intense heat that boiled away its ices into space, to generate a glowing head (the 'coma') and a long tail. Comet Tsuchinshan–ATLAS rounded the Sun on 27 September 2024, and in October it was a brilliant sight in the evening sky as it headed back into deep space. I was treated to a couple of magical nights viewing the comet with the naked eye from North Carolina, though many astronomers in the UK and Ireland were initially frustrated by cloud. But, as you can see in this lovely image from Stuart Atkinson taken later that month, Comet Tsuchinshan–ATLAS continued to be a beautiful sight as seen with binoculars or a telescope for weeks afterwards.

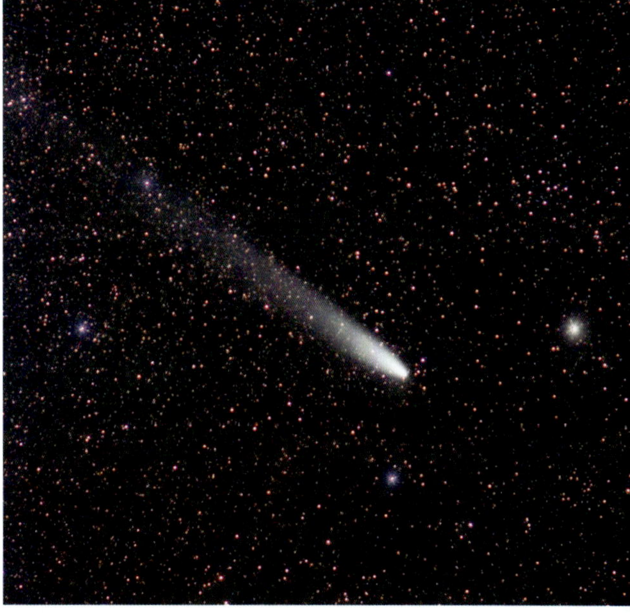

Under the superb sky conditions of the Kielder Forest in Northumberland, Stuart Atkinson captured Comet Tsuchinshan–ATLAS on 29 October 2024 with a Canon EOS 700D DSLR and a 135-mm lens, tracking the sky on an iOptron SkyTracker. It's a crop from a stack of 30 images.

MAY'S CALENDAR

SUNDAY	MONDAY	TUESDAY	WEDNESDAY	THURSDAY	FRIDAY	SATURDAY
31 9.45 am Full Moon; blue Moon					1 6.23 pm Full Moon	2
3 Moon near Antares	4 Moon near Antares (am)	5	6 Eta Aquarids (am)	7	8	9 10.10 pm Last Quarter Moon
10	11	12	13	14 Moon near Saturn (am)	15	16 9.01 pm New Moon
17	18 Moon near Venus	19 Moon between Venus and Jupiter	20 Moon near Jupiter	21	22 Moon near Regulus	23 12.11 pm First Quarter Moon
24	25	26 Moon near Spica	27 Moon near Spica	28	29	30 Moon near Antares

SPECIAL EVENTS

- **Night of 3/4 May:** the Moon glides below Antares.
- **6 May, early hours:** the maximum of the Eta Aquarid meteor shower is spoilt this year by bright moonlight.
- **14 May, 4.15 am:** look to the lower right of the Moon, just above the horizon, to spot Saturn reappearing in the morning sky.
- **18 May:** the crescent Moon and Venus form a lovely duo in the north-west after sunset, with Jupiter to the upper left (Chart 5a).
- **19 May:** the three brightest objects in the night sky converge in Gemini, as the Moon lies between Jupiter (left) and Venus (lower right), with Castor and Pollux above them (Chart 5a).
- **20 May:** you'll find Jupiter just below the crescent Moon, with Castor and Pollux to the right and Venus well down to the lower right (Chart 5a).
- **31 May:** the month's second Full Moon is a blue Moon (see Topic).

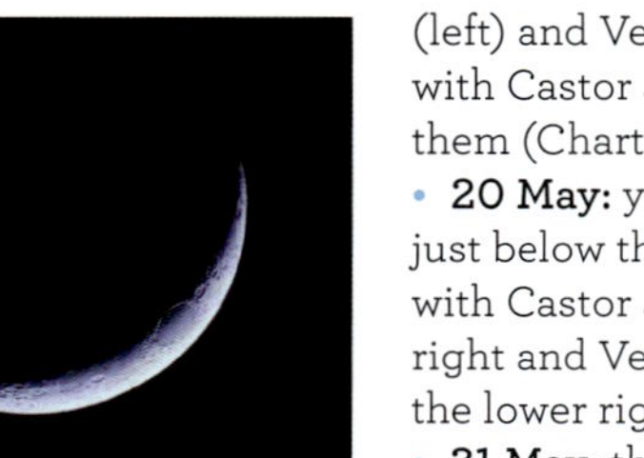

Crescent Moon

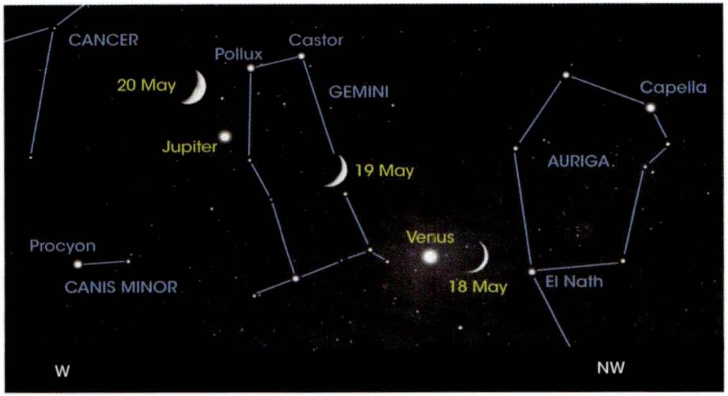

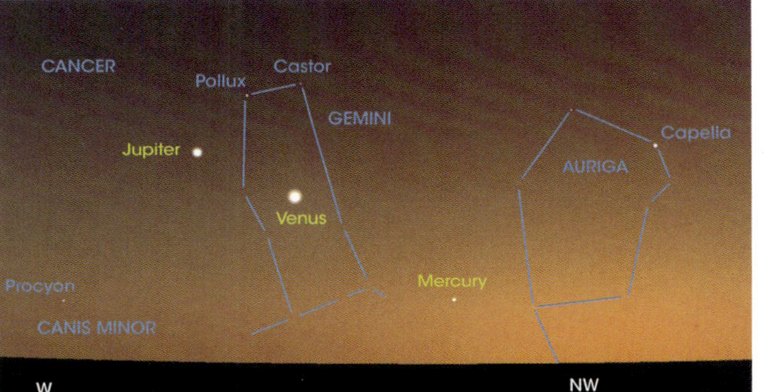

5a 18–20 May, 10.30 pm. The Moon passes Venus and Jupiter.

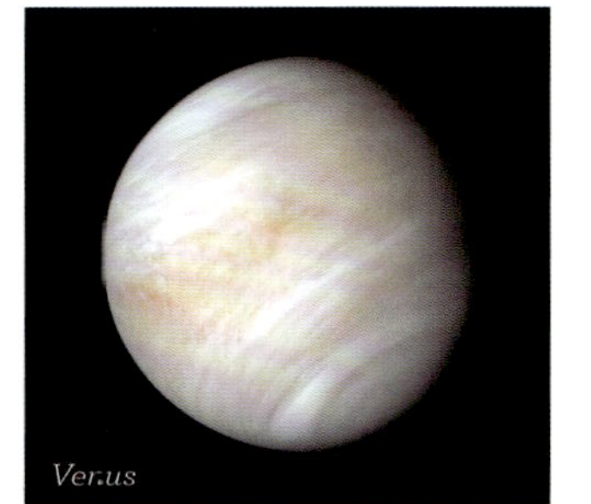

5b 31 May, 10 pm. Mercury joins Venus and Jupiter.

- **Venus** is at its most impressive this month, blazing at magnitude –3.9 in a totally dark sky and – by the end of May – setting as late as midnight. The crescent Moon lies near the Evening Star on 18 and 19 May (Chart 5a). During the last few days of May, look out for Venus inexorably homing in on Jupiter, in preparation for a close conjunction next month.

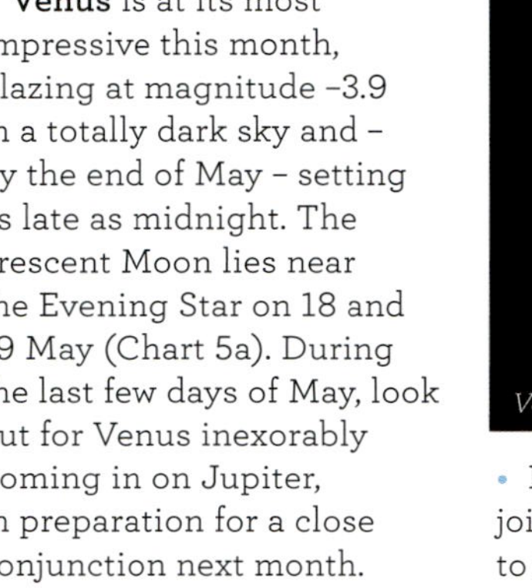

Venus

- From 20 May, Venus is joined by **Mercury**, well down to its lower right.

By the end of the month, the innermost planet is shining at magnitude –0.6 and setting around 11 pm (Chart 5b).

- **Jupiter** (magnitude –1.9) lies in Gemini, to the lower left of the constellation's twin bright stars, Castor and Pollux. The giant planet sets about 1 am. The Moon lies to the right of the giant planet Jupiter on 19 May, and to its upper left on 20 May (Chart 5a).

- In the middle of May, **Saturn** begins to emerge from the twilight into the dawn sky in Pisces. It lies to the lower right of the crescent Moon on the morning of 14 May. By the end of the month, the ringworld is rising around 3 am.

- **Mars**, **Uranus** and **Neptune** are too close to the Sun to be seen this month.

MAY'S PLANET WATCH

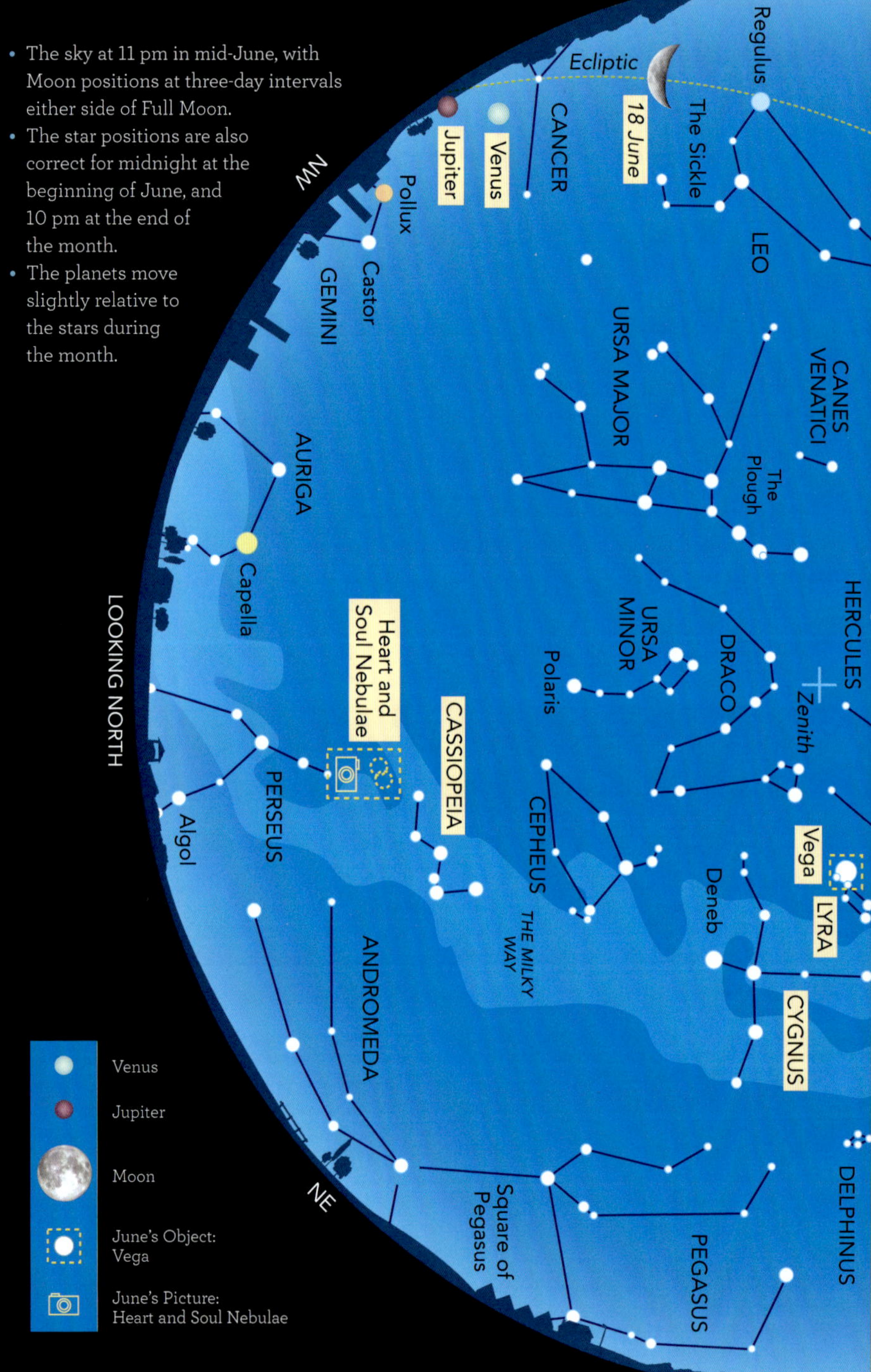

- The sky at 11 pm in mid-June, with Moon positions at three-day intervals either side of Full Moon.
- The star positions are also correct for midnight at the beginning of June, and 10 pm at the end of the month.
- The planets move slightly relative to the stars during the month.

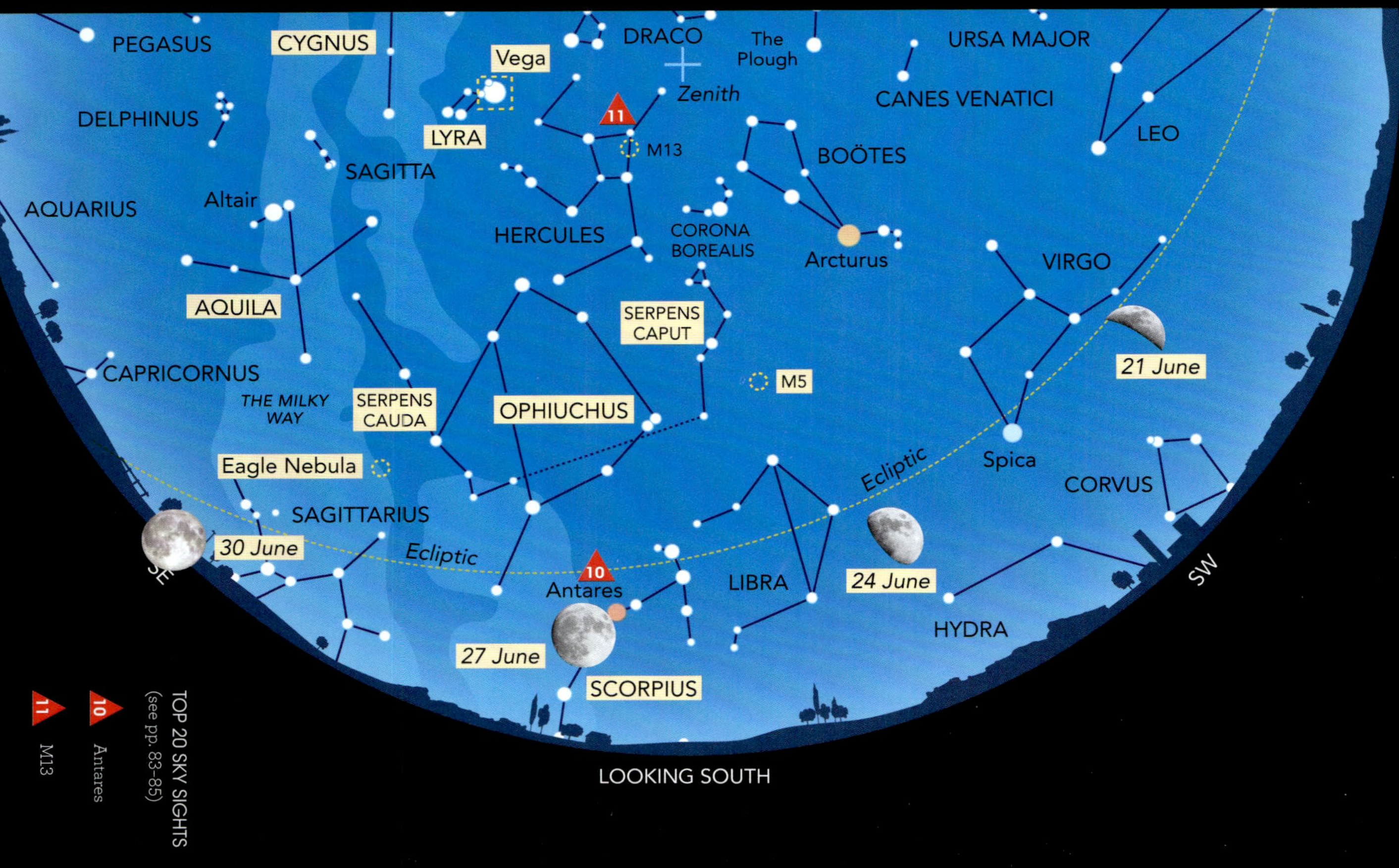
EAST
WEST
PEGASUS
CYGNUS
Vega
DRACO
The Plough
URSA MAJOR
Zenith
CANES VENATICI
DELPHINUS
LYRA
11
M13
BOÖTES
LEO
SAGITTA
AQUARIUS
Altair
HERCULES
CORONA BOREALIS
Arcturus
VIRGO
AQUILA
SERPENS CAPUT
21 June
CAPRICORNUS
M5
THE MILKY WAY
SERPENS CAUDA
OPHIUCHUS
Ecliptic
Spica
CORVUS
Eagle Nebula
SAGITTARIUS
24 June
SE
30 June
Ecliptic
10
Antares
LIBRA
SW
27 June
HYDRA
SCORPIUS
LOOKING SOUTH
TOP 20 SKY SIGHTS
(see pp. 83–85)
10   Antares
11   M13
JUNE

We're treated to the wonderful sight of the two brightest planets together, with **Venus** closest to **Jupiter** on 9 June. And take advantage of the warmer summer nights to enjoy the lovely summer constellations of **Scorpius**, **Lyra**, **Cygnus** and **Aquila**.

JUNE'S CONSTELLATION

We have a unique pair of intertwined star patterns this month: **Ophiuchus** (the Serpent Bearer) and **Serpens** (the Serpent), representing a man gripping a snake. In Roman mythology, the serpent bearer was Aesculapius who – after seeing a serpent cure a mortally wounded snake with a herb – used the same plant to save human lives. But these star patterns are probably millennia older, when the snake's vertical neck represented the 'Greenwich meridian' of the sky.

Though Ophiuchus is the larger, Serpens is the more interesting star pattern. For a start, it's the only constellation that comes in two unconnected parts. **Serpens Caput** (the snake's head) lies to the right of Ophiuchus, with **Serpens Cauda** (its tail) to the left. In Serpens Caput, binoculars reveal **M5**, one of the oldest star clusters in the Milky Way and a splendid sight in a moderate telescope. And Serpens Cauda is home to the **Eagle Nebula**, a giant cloud of glowing gas speckled with newborn stars. Nestled in the centre are dark towering silhouettes of dust and gas, which achieved fame as the Pillars of Creation, after they appeared in a stunning image from the Hubble Space Telescope.

JUNE'S OBJECT

Beautiful **Vega**, in **Lyra**, soars overhead during the summer months – fittingly, as this star's name means 'the swooping eagle'. Because the Earth's axis slowly changes its direction in space (see May's Constellation), Vega was our Pole Star around 12,000 BC and will be again in AD 13,700.

Sara Wager captured the Heart and Soul with a Takahashi FSQ-85 85-mm refractor using an Atik 460EX monochrome camera: it's such a large target that she had to make a six-pane mosaic, involving multiple exposures at two wavelengths – 70 × 1800 seconds in hydrogen alpha and 71 × 1800 seconds in oxygen light – for a total exposure of 70.5 hours.

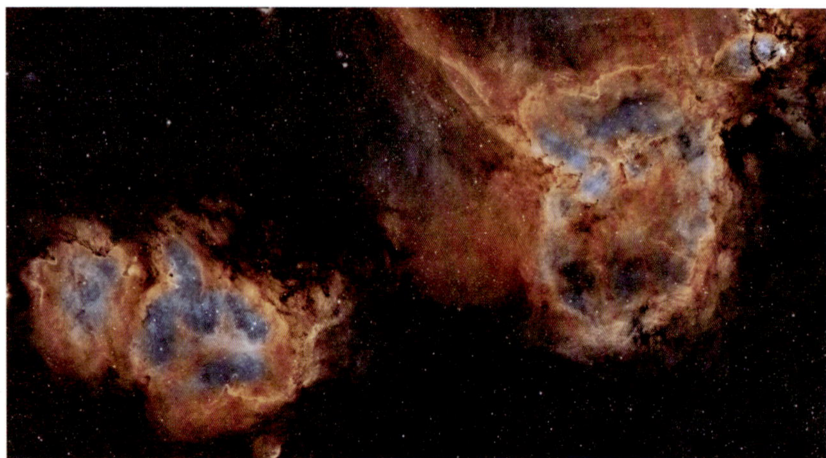

The fifth-brightest star in the sky, Vega was the first star (after the Sun) to be photographed and to have its spectrum recorded. Its colour is so pure white that Vega's hue is used as a benchmark to measure the colours of other stars and so gauge their temperatures (see January's Topic).

Vega was one of the first stars around which astronomers discovered a warm, dusty disc. A ring of dust around a very young star often comprises material forming into new planets, but Vega is a more mature 700 million years old. Here we're probably seeing the debris from collisions between large asteroids or comets – but, so far, no planets have been discovered orbiting Vega.

JUNE'S TOPIC: DARK MATTER

We think of the Universe as being luminous: alight with stars, nebulae and galaxies. But nothing could be further from the truth. Astronomers have discovered that 85 per cent of the matter in the Cosmos is invisible – taking the form of mysterious 'dark matter'.

Suspicions were first aroused in 1933, when Swiss astronomer Fritz Zwicky found that galaxies in the Coma Cluster were moving unexpectedly fast. Something was exerting more gravity than the galaxies in the cluster, and Zwicky called it 'dark matter'.

Later, American researcher Vera Rubin found that stars and gas in the outer regions of spiral galaxies were spinning round much faster than predicted. Again, dark matter in the galaxy must be holding them in.

Most scientists think dark matter is composed of a vast sea of subatomic particles filling all of space, but concentrated

in galaxies and galaxy clusters. That means that dark matter is everywhere, including inside you and me!

Experiments around the world are now trying to catch dark matter particles and identify what they are. So far there are no definitive answers, but the solution to this almost-century-old cosmological mystery could come at any time from a lab right here on Earth.

JUNE'S PICTURE

If your heart and soul are into astronomy, then this month's pair of objects is a must for you! The **Heart** and **Soul Nebulae** are neighbours in Cassiopeia, around 7500 light years away.

The Heart Nebula (right) is named after its characteristic shape. Some 300 light years across, the nebula is centred on a cluster of massive young stars, only 2 million years old. In the bright knot at the top right (known as the Fish Head Nebula) young stars are being born right now. The Soul Nebula contains several small clusters, 3–5 million years old.

As photographer Sara Wager puts it: 'The two nebulae are both massive star-making factories, marked by giant bubbles blown into surrounding dust by radiation and winds from the stars.'

JUNE'S CALENDAR

SUNDAY	MONDAY	TUESDAY	WEDNESDAY	THURSDAY	FRIDAY	SATURDAY
	1	2	3	4	5	6
7	8 11.00 am Last Quarter Moon	9 Venus near Jupiter	10 Moon near Saturn (am)	11	12	13 Moon near Mars (am)
14	15 3.54 am New Moon; Mercury E elongation	16 Moon near Mercury and Jupiter	17 Moon very near Venus	18	19 Venus near Praesepe; Moon near Regulus	20
21 10.55 pm First Quarter Moon; Summer Solstice	22	23 Moon near Spica	24	25	26	27 Moon near Antares
28	29 Mars near Pleiades	30 0.56 am Full Moon				

SPECIAL EVENTS

- **9 June:** a splendid conjunction of the two most brilliant planets, as Venus passes just 1.5 degrees from Jupiter, with Castor and Pollux to the right and Mercury to the lower right (Chart 6a).
- **10 June, before dawn:** Saturn lies below the crescent Moon.
- **13 June, before dawn:** look low on the horizon to the right of the crescent Moon to spot the Red Planet, Mars.
- **15 June:** Mercury at its greatest separation from the Sun.
- **16 June:** Mercury lies immediately below the crescent Moon (best seen in binoculars), with Venus and Jupiter to the upper left (Chart 6b).
- **17 June:** Venus and the crescent Moon form a dazzling duo in the evening sky after sunset, less than a degree apart. Jupiter lies to the lower right, with Mercury further down on the horizon (Chart 6b).
- **19 June:** Venus passes just above Praesepe.
- **21 June, 9.24 am:** Summer Solstice. The Sun reaches its most northerly point, so today is Midsummer's Day, with the longest day and the shortest night.
- **29 June:** Mars is 4 degrees below the Pleiades.

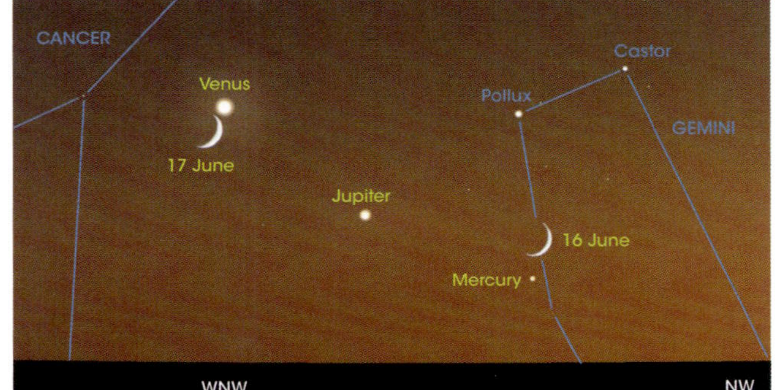

6a 9 June, 10.30 pm. Conjunction of Venus and Jupiter, with Mercury nearby.

6b 16–17 June, 10.30 pm. The Moon passes Mercury, Jupiter and very close to Venus.

- Staying above the horizon until midnight, **Venus** is glorious in the west at magnitude –4.0. Moving inexorably upwards during the month, the Evening Star passes only 1.5 degree from Jupiter on 9 June (see Special Events and Chart 6a). The crescent Moon is very near Venus on 17 June (Chart 6b). On 19 June, you'll find Venus just above the pretty star cluster Praesepe – a lovely sight in binoculars or a low-power telescope.
- **Jupiter** is in the same neck of the woods as the Evening Star; some eight times fainter than Venus at magnitude –1.8. it's still brighter than any of the stars. The giant planet lies in Gemini, in line with the constellation's main stars, Castor and Pollux, and sets about 11 pm.
- During the first half of June, you can find **Mercury** just above the horizon, to the lower right of Venus and Jupiter, and at its greatest elongation on 15 June. Setting around 11 pm, the innermost planet fades from magnitude –0.5 on 1 June to magnitude +0.6 on 16 June, when the crescent Moon lies right above Mercury (Chart 6b).
- **Saturn** is now prominent in the morning sky, rising about 2 am and shining at magnitude +0.8 on the border of Pisces and Cetus. The Moon is nearby on the morning of 10 June.
- **Neptune** (magnitude +7.8) lies to the right of Saturn, in Pisces, rising around 1.30 am.
- Mid-month, **Mars** appears in the north-eastern dawn twilight (about 3 am) in Taurus. It passes below the Pleiades on 29 June (see Special Events).
- **Uranus** is lost in the Sun's glare in June.

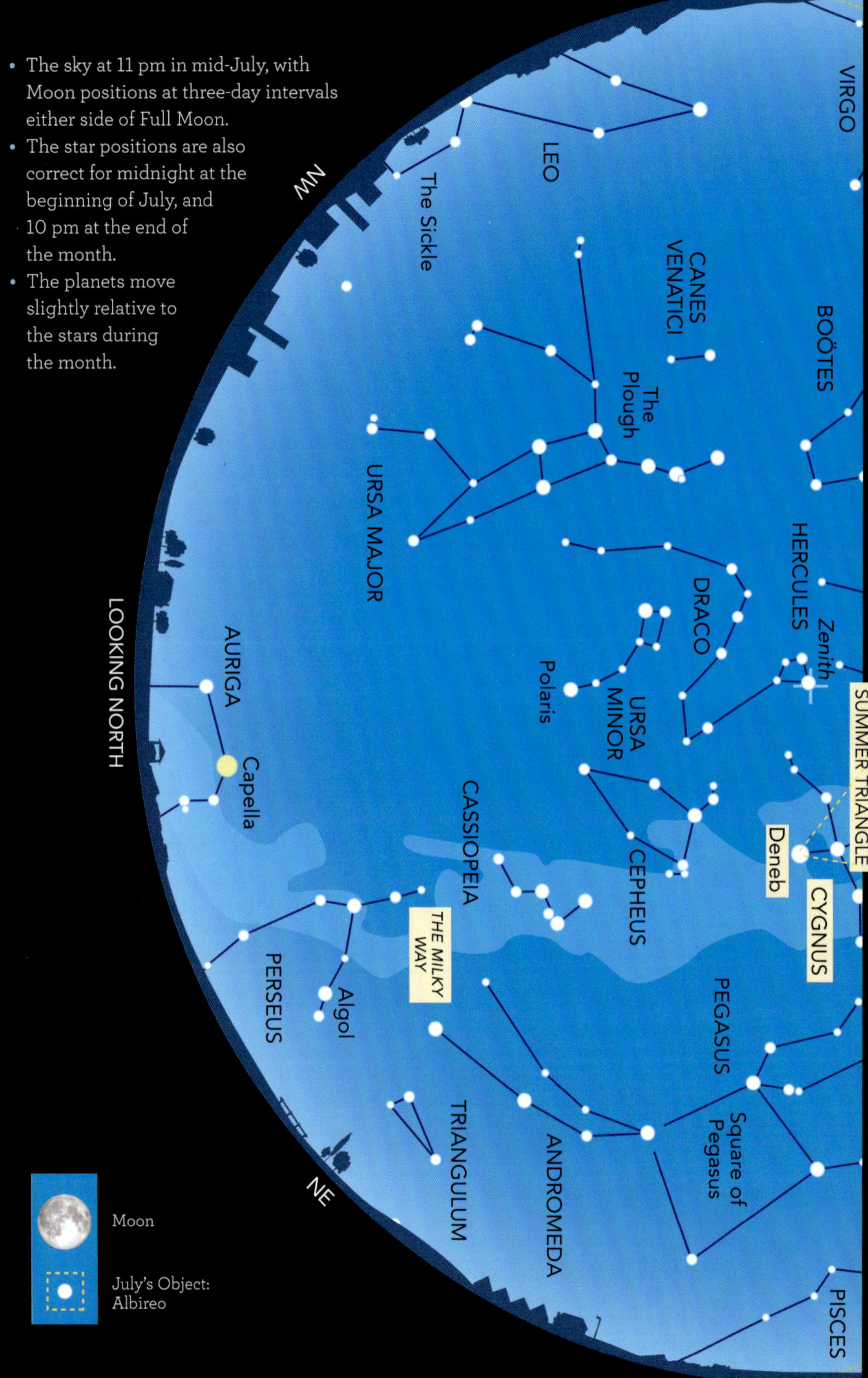

- The sky at 11 pm in mid-July, with Moon positions at three-day intervals either side of Full Moon.
- The star positions are also correct for midnight at the beginning of July, and 10 pm at the end of the month.
- The planets move slightly relative to the stars during the month.

   JULY

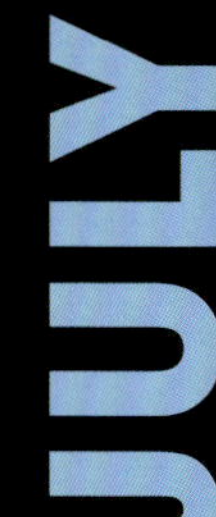

JULY 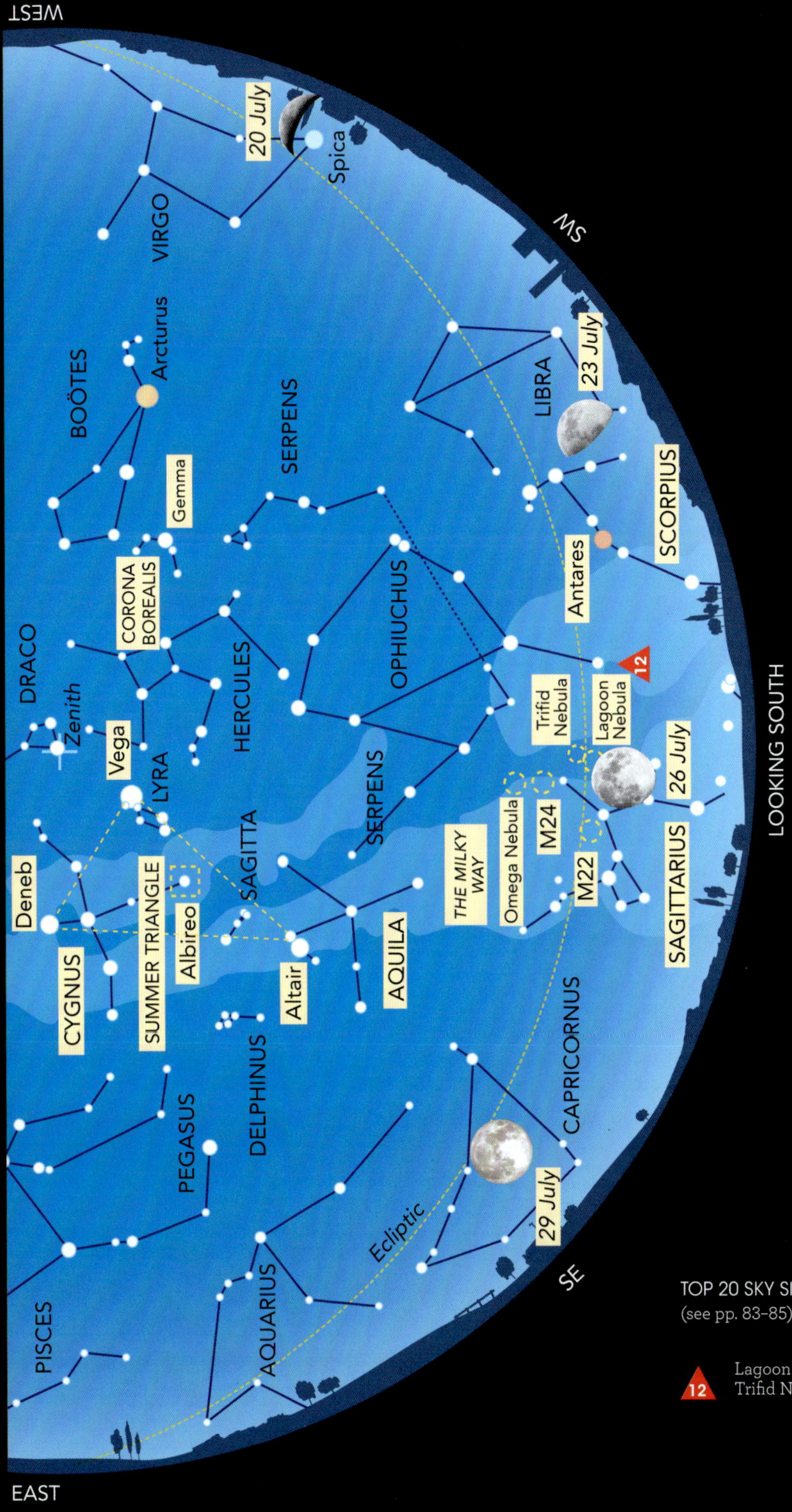
WEST
VIRGO
BOÖTES
Arcturus
Gemma
CORONA BOREALIS
DRACO
Zenith
Vega
LYRA
HERCULES
SERPENS
OPHIUCHUS
SERPENS
Spica
20 July
MS
LIBRA
23 July
SCORPIUS
Antares
Trifid Nebula
Lagoon Nebula
12
M24
M22
Omega Nebula
THE MILKY WAY
SAGITTA
Albireo
SUMMER TRIANGLE
Deneb
CYGNUS
AQUILA
Altair
DELPHINUS
PEGASUS
AQUARIUS
PISCES
EAST
Ecliptic
29 July
SE
CAPRICORNUS
SAGITTARIUS
26 July
LOOKING SOUTH
JULY
45
TOP 20 SKY SIGHTS
(see pp. 83–85)
12
Lagoon and
Trifid Nebulae

Venus is the queen of the summer nights, with most of the other planetary action taking place in the early hours. Two ancient constellations – **Sagittarius** and **Scorpius** – decorate the southern region of the sky. Higher up, the brilliant stars **Vega**, **Deneb** and **Altair** mark out the vertices of the **Summer Triangle**.

JULY'S CONSTELLATION

To the ancient Greeks, the star pattern **Sagittarius** represented an archer, with the torso of a man and the body of a horse, though to modern eyes the shape suggests merely a teapot.

Sagittarius is rich in nebulae and star clusters, easily visible in binoculars. Above the spout lies the wonderful **Lagoon Nebula** – a region of starbirth that's visible to the naked eye on a really dark night. Its neighbour, the three-lobed **Trifid Nebula**, requires a telescope.

Between Sagittarius and **Aquila**, you'll find a bright patch of stars in the Milky Way, catalogued as **M24**. Raise your binoculars higher to spot another star-forming region, the **Omega Nebula**. Finally, the fuzzy patch **M22** is a globular cluster of almost a million stars, lying 11,000 light years away.

JULY'S OBJECT

Albireo marks the head of the constellation **Cygnus**, a swan with outspread wings flying down the **Milky Way**. Though it may look unassuming to the unaided eye, turn even the smallest telescope on Albireo and you'll find it's a double star – and one of the most glorious sights in the night sky. Some 360 light years away, a dazzling yellow star is teamed up with a blue companion.

The yellow star is a giant, near the end of its life. It's 60 times bigger than the Sun, and 1200 times brighter. The fainter blue companion is 'only' 230 times brighter than the Sun.

The spectacular colour contrast is due to the stars' different temperatures (see January's Topic). The giant star is slightly cooler than our Sun, and has a similar hue. The smaller companion has such a high temperature that it shines not merely white hot, but blue-white.

JULY'S TOPIC:
MAYAN ASTRONOMY

Over a thousand years ago, astronomy was thriving in the humid jungles of what are now Mexico and Guatemala – and it was largely based on Venus. The Mayan people built large stone 'observatories' that were aligned with the planet's rising and setting points. The few fragments of their books that have survived show that the Mayan astronomers were well aware

OBSERVING TIP

This is the month when you really need a good, unobstructed horizon to the south for the best views of the glorious summer constellations of Scorpius and Sagittarius. They never rise high in temperate latitudes, so make the best of a southerly view – especially over the sea – if you're away on holiday. A good southern horizon is also best for views of the planets, because they rise highest when they're in the south.

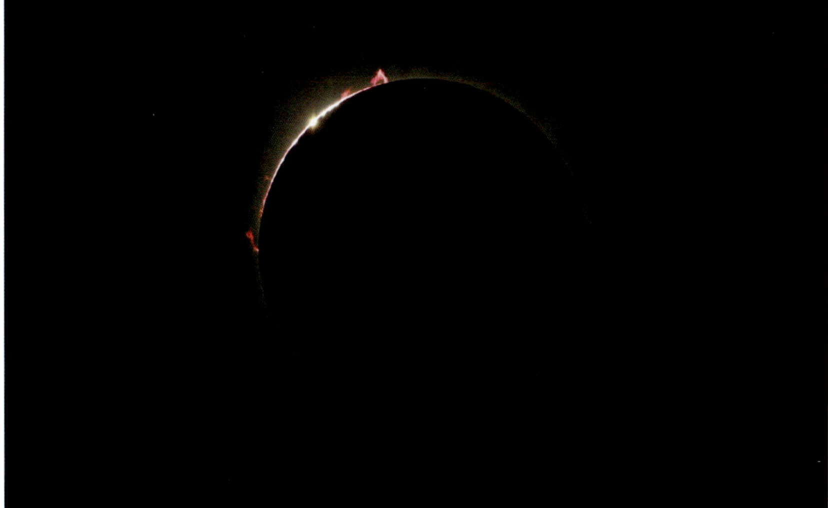

that the position of Venus in the sky repeats almost exactly every eight years. As a result, they could accurately predict the planet's motions up to 500 years in the future!

They also devised a complex calendar system, based on a 'week' of 20 named days. The sacred calendar had 13 weeks, so it repeated every 260 days. For practical purposes, they used a calendar of 18 weeks plus five odd days, for a total of 365 days. The two calendars lined up every 52 years, when the Mayans held special celebrations.

They kept careful track of the seasons, too. A stepped pyramid at Chichén Itzá, Mexico, is aligned so that – at the equinoxes – the zigzag shadow of its nine levels moves down the side of a processional staircase, like a writhing snake, towards a carved snake's head at the base of the staircase.

JULY'S PICTURE

Astronomers across Europe are preparing for next month's solar eclipse. From these islands, it will be the biggest partial eclipse since the last total solar eclipse back in 1999. Or book a ticket, pack your

Among the many images Martin Ratcliffe took of the eclipsed Sun, this one had a relatively short exposure to show the bright chromosphere rather than the tendrils of the extensive faint corona (the outer atmosphere). It was taken with a Canon EOS Rebel T5i camera (internal infrared filter removed) on a Stellarvue SV80 80-mm refractor, with a shutter speed of 1/4000 second at ISO 200.

bags and fly to Spain or Iceland to view totality (see August's Topic).

Total solar eclipses are not rare, though, if you're prepared to go far enough. Somewhere in the world there's a total solar eclipse every year or two, and Martin Ratcliffe captured this splendid image of the eclipsed Sun as recently as 2024 from the United States. He stationed himself at East Tawakoni in Texas on 8 April, where he was treated to 4 minutes 21 seconds of totality.

This eclipse was memorable for towering prominences in the Sun's lower atmosphere, the chromosphere, glowing pink in the light from hydrogen gas suspended in giant magnetic loops. Adding to Martin's mesmerising shot is the gleam of the Sun's brilliant surface shining through a valley in the Moon's surface, creating the aptly named 'diamond ring'.

JULY'S CALENDAR

SUNDAY	MONDAY	TUESDAY	WEDNESDAY	THURSDAY	FRIDAY	SATURDAY
			1	2	3	4 Mars very near Uranus (am)
5	6 Earth at aphelion	7 8.29 pm Last Quarter Moon; Moon near Saturn (am)	8 Moon near Saturn (am)	9 Venus near Regulus	10	11 Moon near the Pleiades and Mars (am)
12 Moon near Mars (am)	13	14 10.43 am New Moon	15	16 Moon near Regulus	17 Moon near Venus	18
19	20 Moon near Spica	21 12.05 pm First Quarter Moon	22 Moon near Jupiter (am)	23	24 Moon near Antares	25
26	27	28	29 3.36 pm Full Moon	30	31	

SPECIAL EVENTS

- **4 July, am:** Mars passes very close to Uranus (see Planet Watch and Chart 7a).
- **6 July, 6.30 pm:** the Earth is furthest from the Sun (aphelion), at 152 million km.
- **7 July, am:** Saturn lies to the lower left of the Moon.
- **8 July, am:** the Moon is to the upper left of Saturn.
- **9 July:** Venus passes less than a degree from Regulus.
- **11 July, am:** as you watch the Moon rising, around 1 am, you'll notice that it's right beside the Pleiades, with Mars to the lower left.
- **12 July, am:** the crescent Moon lies to the left of Mars.
- **17 July:** the crescent Moon forms a lovely duo with Venus low in the west after sunset, with Regulus to the right (Chart 7b).
- **24 July:** enjoy a lovely summer evening's sight, as the Moon glides under Antares low in the southern sky.

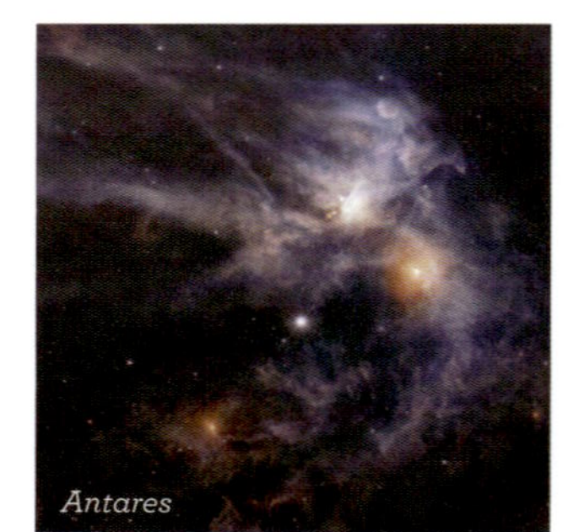
Antares

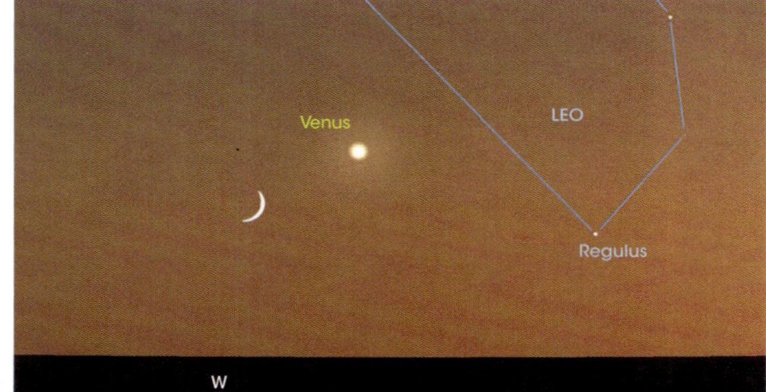

7a *4 July, 3 am. Mars passes close to Uranus (binocular view).*

7b *17 July, 10 pm. The Moon near to Venus and Regulus.*

- **Venus** is the 'star' of the evening sky, blazing at magnitude –4.2 over in the west and setting about 11 pm. On 9 June, the Evening Star passes right above Regulus. The Moon lies to the left of Venus on 17 June (Chart 7b).
- At the start of July, you'll find **Jupiter** well down to the lower right of Venus in the twilight haze. Lying in Cancer, it shines at magnitude –1.8 and sets at 10.30 pm. But after a few days the giant planet sinks out of sight into the Sun's glow.
- The rest of the planetary action takes place in the wee small hours. The first planet you'll notice is **Saturn**, shining at magnitude +0.8 in Pisces and rising just after midnight. The Moon is nearby on 7 and 8 July.
- **Neptune** lies about 10 degrees to the right of Saturn, also in Pisces and rising about half an hour earlier. At magnitude +7.7, you'll need binoculars or a telescope to spot the eighth planet.
- **Mars** (magnitude +1.3) rises around 2 am in Taurus. During July it tracks from a position near the Pleiades, past Aldebaran on 13 July to reach the tip of the Bull's 'horns' – to the extreme left of the constellation – by the end of the month.
- Also in Taurus, near the Pleiades, lies **Uranus**. The seventh planet rises about 1.30 am; shining at a dim magnitude +5.8, it's best seen in binoculars. But you have an excellent chance to spot Uranus on the morning of 4 July, when Mars – 60 times brighter – sails by at a distance of only 9 arcminutes. It's your opportunity to view the reddest and the greenest planets of the Solar System in the same binocular or telescope field of view (Chart 7a)!
- **Mercury** is too close to the Sun to be visible in July.

- The sky at 11 pm in mid-August, with Moon positions at three-day intervals either side of Full Moon.
- The star positions are also correct for midnight at the beginning of August, and 10 pm at the end of the month.
- The planets move slightly relative to the stars during the month.

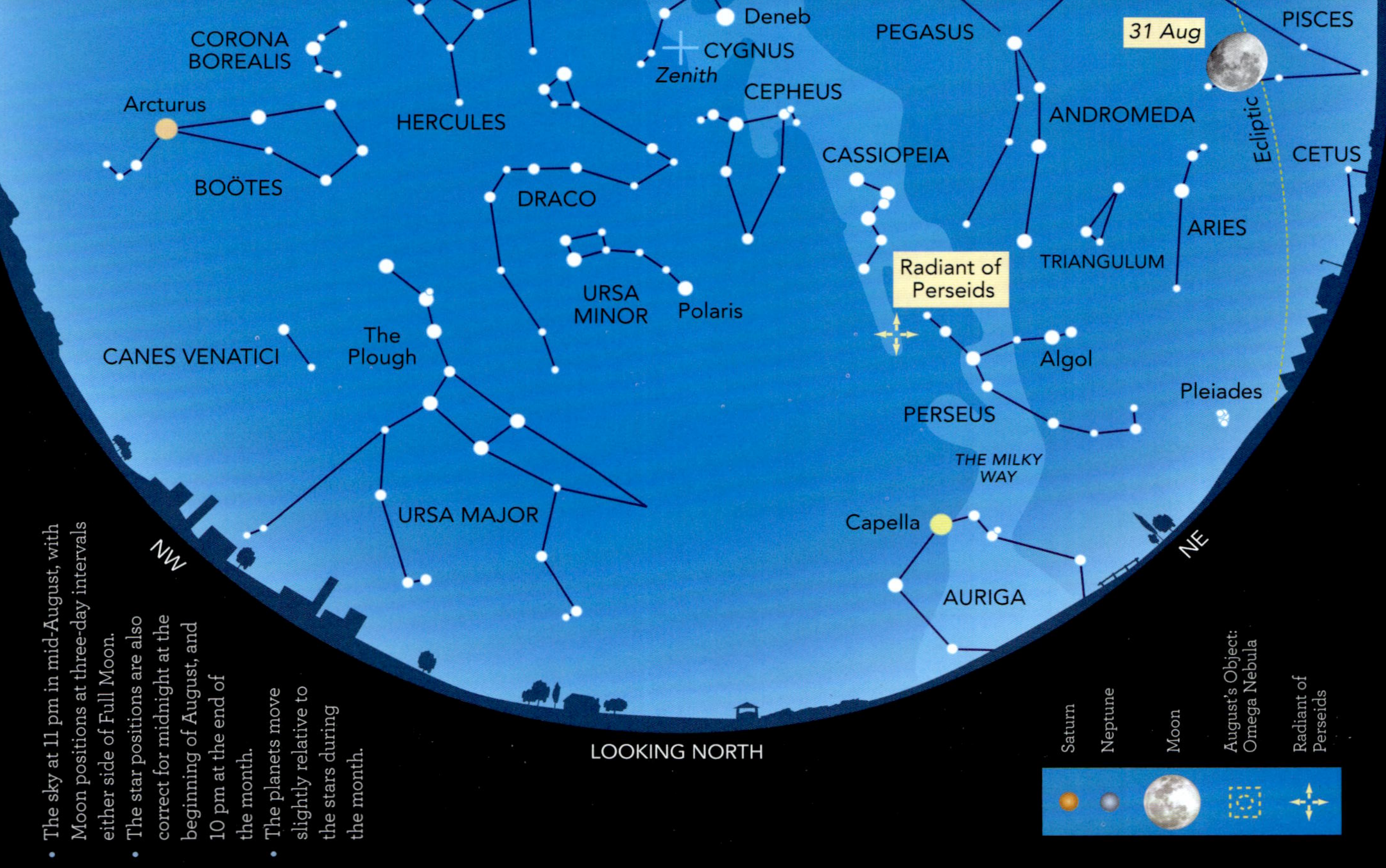

EAST
WEST
31 Aug
ANDROMEDA
DRACO
Zenith
CORONA BOREALIS
PISCES
Deneb
Double-Double
Vega
Square of Pegasus
LYRA
13
Saturn
CYGNUS
Ring Nebula
HERCULES
Albireo
SERPENS
Neptune
Dumbbell Nebula
14
PEGASUS
SAGITTA
LIBRA
DELPHINUS
Altair
SERPENS
OPHIUCHUS
CETUS
28 Aug
AQUARIUS
AQUILA
SCORPIUS
Eagle Nebula
Omega Nebula
SE
Ecliptic
25 Aug
SW
PISCIS AUSTRINUS
22 Aug
CAPRICORNUS
SAGITTARIUS
LOOKING SOUTH

TOP 20 SKY SIGHTS
(see pp. 83–85)

13 Albireo

14 Dumbbell Nebula

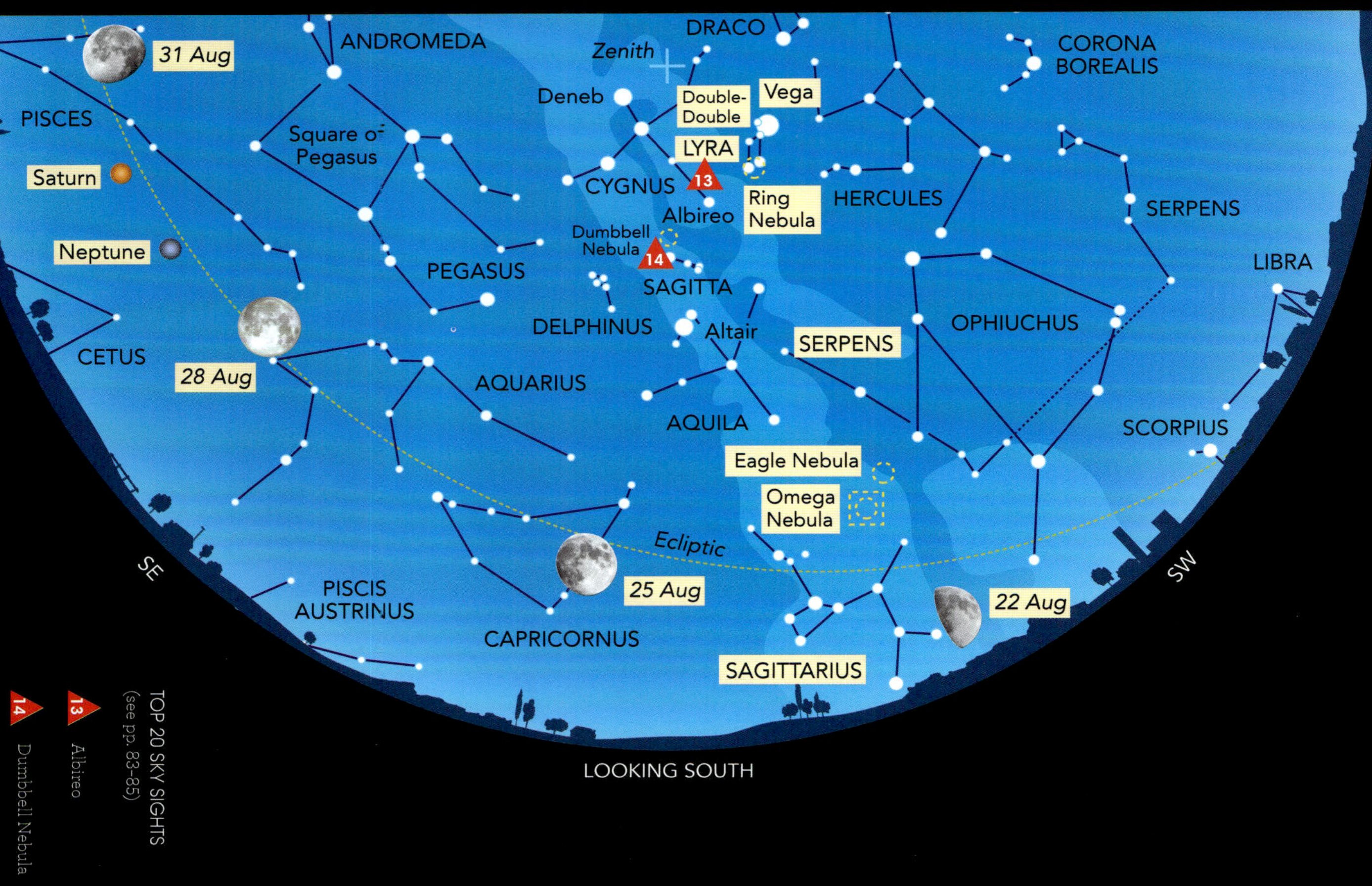

AUGUST

It's eclipse month! As seen from the UK and Ireland, both the Sun and the Moon suffer large partial eclipses. Travel to Iceland, Spain or the eastern Atlantic Ocean, however, and you'll be treated to the glorious spectacle of a total solar eclipse. As a bonus, that same night we're treated to a display of cosmic fireworks.

AUGUST'S CONSTELLATION

Lyra is small but perfectly formed. Named after the ancient Greek musical instrument, the lyre, it's dominated by brilliant white **Vega**, the fifth-brightest star in the sky. Just 25 light years away – a near neighbour in the Cosmos – Vega is surrounded by a disc of dust that's probably debris from asteroids or comets that have collided.

Next to Vega is the **Double-Double**, a quadruple star known officially as epsilon Lyrae. Keen-sighted people can see a pair of stars here, and a telescope reveals that each star is itself double.

A fainter gem of Lyra lies between the two end stars of the constellation. Visible in a small telescope, the **Ring Nebula** is a wonderful example of a planetary nebula: the ghostly remains of a star that puffed away its atmosphere in its death throes, some 1600 years ago.

AUGUST'S OBJECT

The **Omega Nebula** – in **Sagittarius** – is an overlooked twin. Close by in **Serpens**, the **Eagle Nebula** is famous because it contains the iconic Pillars of Creation snapped by the Hubble Space Telescope. The Omega Nebula is equally bright (magnitude +6.0), and a striking sight in a small telescope.

In 1823, John Herschel compared the nebula to the Greek letter Omega (Ω), though only one horizontal bar is visible and the loop is faint. Charles Messier, listing the nebula as number 17 in his 1781 catalogue, saw only the bar, and described it as 'a train of light without stars'.

And that's what makes the Omega Nebula so unusual. When we look at most nebulae (such as the Orion or Lagoon Nebulae), they appear bright largely because our eyes are also taking in light from a central cluster of stars. While there is a star cluster associated with the Omega, it's hidden by a dark dust cloud within the loop. So the Omega Nebula is among the brightest clouds of purely luminous gas that we can see in the sky.

AUGUST'S TOPIC: TOTAL ECLIPSE

Eclipse fever is building in Europe this month as the continent is about to experience its first total solar eclipse in 27 years. On 12 August, the shadow of

OBSERVING TIP

Whether you're observing the partial solar eclipse this month, or just checking out our local star, NEVER look at the Sun directly with your unprotected eyes or – especially – with a telescope or binoculars: it can blind you permanently. For naked-eye observing, use special 'eclipse glasses' with dark filters (meeting the ISO 12312-2 safety standard). Or you can attach a sheet of a specialised metallised film (such as Baader AstroSolar® film) across the front of your instrument as a safe filter.

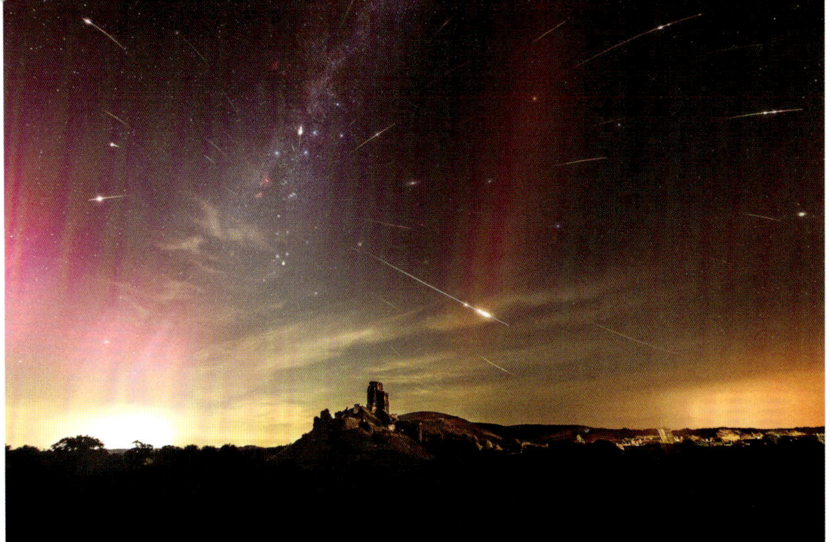

Setting up his Sony Alpha 7S II camera on a tripod to view the sky at Corfe Castle in Dorset, Josh Dury secured this wide-angle view using a Sigma 15-mm f/1.4 DG DN Diagonal Fisheye lens, taking 10-second exposures at ISO 640.
He blended 45 images into a single frame layered in Photoshop.

the Moon races from the North Pole southwards across Greenland to western Iceland, where the eclipse has its maximum duration of 2 minutes 18 seconds.

Sweeping down the eastern Atlantic Ocean, the eclipse path makes landfall again in northern Spain (where totality lasts 1 minute 50 seconds) and then across the Mediterranean to Majorca, where the eclipse ends at sunset.

Cities that will be treated to totality include Reykjavik, A Coruña, Bilbao, Zaragoza, Valencia and Palma, plus the outskirts of Madrid. The eclipse is partial over Canada and the north-eastern United States, along with most of Europe, including the UK and Ireland where more than 90 per cent of the Sun is hidden (see Special Events).

During the partial phase, make sure your eyes are protected (see Observing Tip). As the Sun disappears, you'll feel increasingly damp and cold, while the birds stop singing. At the moment of totality, look at the Sun directly, to witness its glowing corona – like a luminous flower with a dark heart – before totality ends in a brilliant diamond ring.

AUGUST'S PICTURE

There's nothing more romantic on your summer holidays than wishing on a shooting star – except for experiencing a whole shower of meteors! And that's what we are treated to on 12 August (see Special Events).

Every year, the Earth runs into a stream of debris shed by Comet Swift-Tuttle. The specks of cosmic dust (about the size of coffee granules) smash into our atmosphere at a speed of 210,000 kilometres per hour, and burn up in the flash of glory that we call a meteor.

Because of perspective, these shooting stars all appear to diverge from the same part of the sky – the 'radiant' – which lies in the constellation Perseus, as you can see in this stunning image that Josh Dury photographed above an 11th-century castle – coincidentally capturing an aurora as well.

AUGUST'S CALENDAR

SUNDAY	MONDAY	TUESDAY	WEDNESDAY	THURSDAY	FRIDAY	SATURDAY
30 Moon near Saturn	31					1
2 Mercury W elongation	3 Moon near Saturn	4	5	6 3.21 am Last Quarter Moon	7 Moon near the Pleiades (am)	8
9 Moon near Mars (am)	10	11 Moon near Mercury (am)	12 6.37 pm New Moon; total solar eclipse; Perseids	13 Perseids (am)	14 Mercury near Praesepe (am)	15 Mercury near Jupiter (am); Mars near M35; Venus E elongation near Moon
16 Moon near Venus	17	18	19	20 3.46 am First Quarter Moon near Antares	21	22
23	24	25	26	27	28 5.18 am Full Moon; partial lunar eclipse	29

SPECIAL EVENTS

- **2 August:** Mercury reaches its greatest morning separation from the Sun.
- **12 August:** a total eclipse of the Sun is visible from western Iceland and northern Spain (see Topic). Ireland and the UK will witness a partial eclipse (Chart 8a): the Sun is 91 per cent eclipsed as seen from London (7.13 pm) and Edinburgh (7.06 pm), rising to 98 per cent for the south-west of Ireland (7.13 pm).
- **Night of 12/13 August:** with the Moon well out of the way eclipsing the Sun, it's a fantastic year for watching the maximum of the **Perseid meteor shower,** high-speed pellets of dust from Comet Swift–Tuttle burning up high above our heads.
- **15 August, am:** Mercury passes close to Jupiter; Mars is near M35 (see Planet Watch).
- **15 August:** Venus reaches its greatest evening separation from the Sun and forms a lovely sight next to the crescent Moon.
- **16 August:** the crescent Moon lies near Venus.
- **28 August, 3.34–6.52 am:** a large partial eclipse of the Moon is visible from the Americas, Africa and Europe, including the UK and Ireland. At maximum (5.14 am) 93 per cent of the Moon is in the Earth's shadow (Chart 8b).

AUGUST'S PLANET WATCH

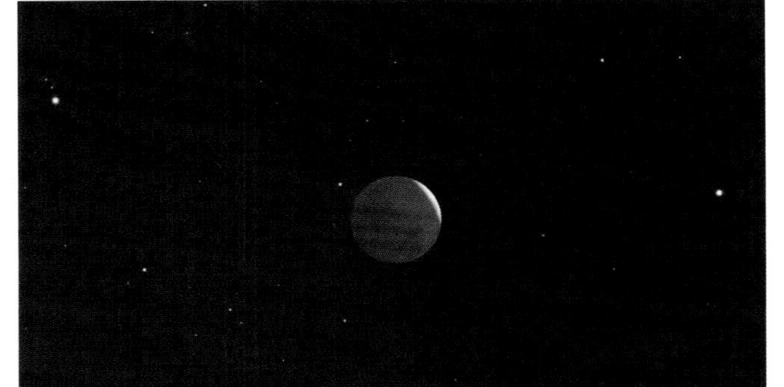

8a *12 August, 7.12 pm (view from Portmagee, Ireland). Partial solar eclipse.*

8b *28 August, 5.14 am. Partial lunar eclipse.*

• Though its brightness is increasing day by day – rising to –4.6 at the end of August – **Venus** is looking less prominent as it sinks down into the twilight glow, setting as early as 9.30 pm. The crescent Moon makes a lovely pairing with Venus on 15 August – when the planet reaches its greatest separation from the Sun – and also on 16 August.

• Soon after the Evening Star sets, **Saturn** rises on the opposite horizon at 10 pm. The ringworld shines at magnitude +0.6 in Pisces. The Moon is nearby on 3 August.

• Also in Pisces, **Neptune** (magnitude +7.7) rises about 9.30 pm, but isn't visible unless you have optical aid.

• Its slightly nearer twin, **Uranus**, shines at magnitude +5.8 in Taurus, and rises around 11.30 pm.

• **Mars** is rising about 1.30 am. Shining at magnitude +1.2, the Red Planet moves from Taurus into Gemini during August. The Moon is nearby on the morning of 9 August. The morning of 15 August sees Mars pass just south of the lovely star cluster M35 (see March's Constellation): a ruby offsetting a gently glowing diadem.

• After its greatest separation from the Sun on 2 August, **Mercury** remains visible low on the horizon for a couple of weeks, rising around 4 am and brightening to magnitude –1.2. On the morning of 14 August, Mercury flies through Praesepe; and in the pre-dawn skies of 15 August, it's only 40 arcminutes from Jupiter – for both events you'll need a clear horizon to the north-east and good binoculars to fight the twilight glow.

• **Jupiter**, in Cancer, reappears in our morning sky mid-month, shining at magnitude –1.8 and rising about 4.30 am.

SEPTEMBER

The sky at 11 pm in mid-September, with Moon positions at three-day intervals either side of Full Moon.
The star positions are also correct for midnight at the beginning of September, and 10 pm at the end of the month.
The planets move slightly relative to the stars during the month.
LOOKING NORTH
NW
NE
EAST
OPHIUCHUS
SERPENS
CORONA BOREALIS
Arcturus
CANES VENATICI
BOÖTES
HERCULES
DRACO
Vega
LYRA
Deneb
CYGNUS
Zenith
The Plough
Mizar
Alcor
URSA MAJOR
URSA MINOR
Polaris
CEPHEUS
15
Delta Cephei
Tsih
CASSIOPEIA
ANDROMEDA
TRIANGULUM
ARIES
29 Sept
Capella
PERSEUS
Algol
AURIGA
THE MILKY WAY
Pleiades
Uranus
Ecliptic
4 Sept
Aldebaran
TAURUS
Saturn
Uranus
Neptune
Moon
September's Object: Delta Cephei
September's Picture: Dumbbell Nebula
Radiant of 45P-ids

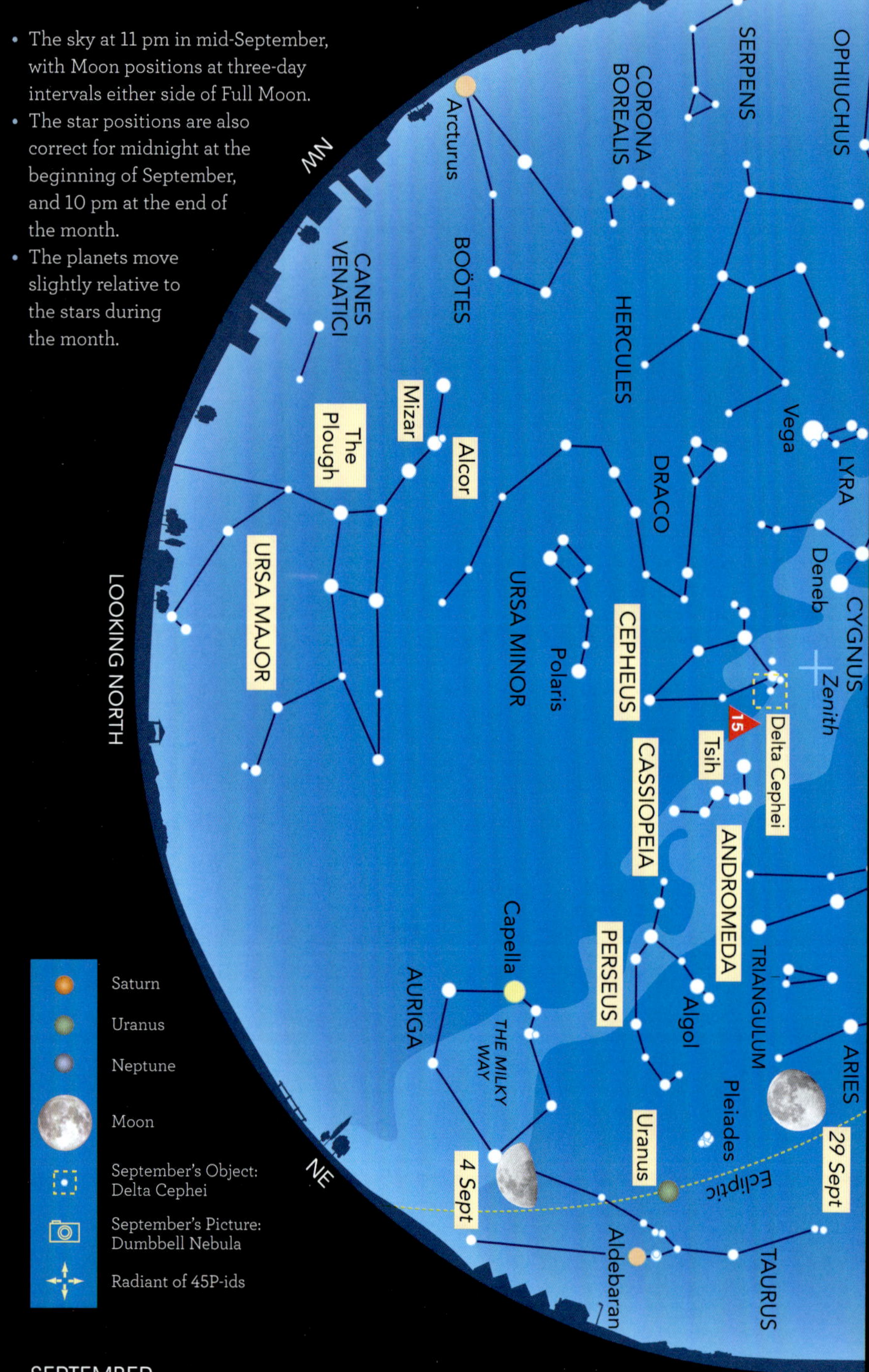

SEPTEMBER

EAST
WEST
TAURUS
29 Sept
ARIES
TRIANGULUM
Delta Cephei
CEPHEUS
Zenith
ANDROMEDA
Deneb
Double-Double
Vega
SERPENS
HERCULES
LYRA
CYGNUS
Albireo
VULPECULA
SAGITTA
OPHIUCHUS
Dumbbell Nebula
ERIDANUS
PISCES
Square of Pegasus
PEGASUS
Altair
SERPENS
DELPHINUS
Saturn
26 Sept
AQUILA
THE MILKY WAY
CETUS
Neptune
Radiant of 45P-ids
Algedi
SAGITTARIUS
AQUARIUS
Ecliptic
23 Sept
20 Sept
SW
SE
Fomalhaut
PISCIS AUSTRINUS
CAPRICORNUS
GRUS
LOOKING SOUTH
TOP 20 SKY SIGHTS
(see pp. 83–85)
15 Delta Cephei
SEPTEMBER
57

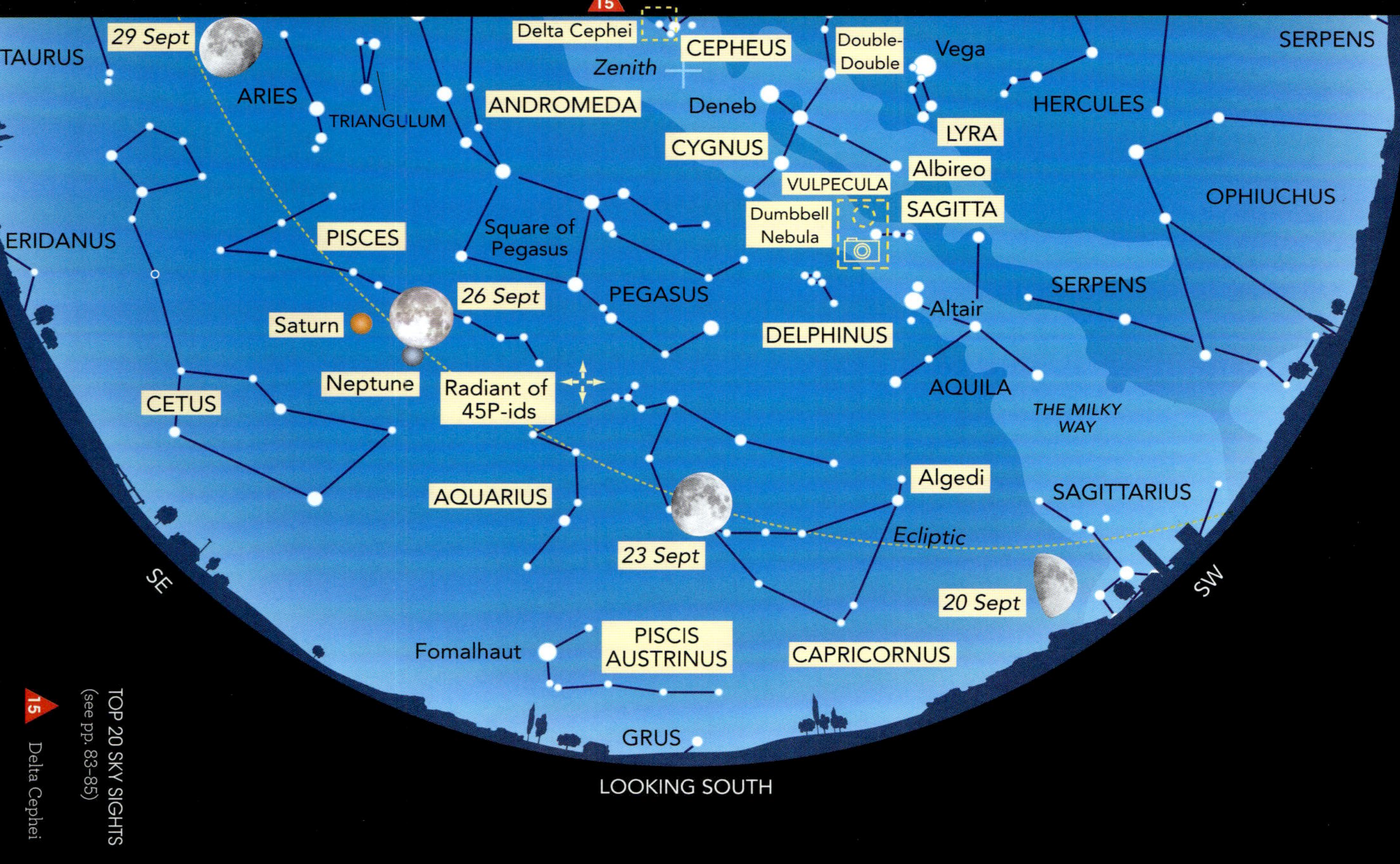

After months as a brilliant Evening Star, Venus drops out of sight this month. On the starry front, the sky is awash with watery constellations: **Piscis Austrinus** (the Southern Fish) and **Aquarius** (the Water Carrier) are joined by **Delphinus** (the Dolphin), the strangely named Sea Goat (**Capricornus**), a pair of Fishes (**Pisces**) and **Cetus** (the Sea Monster).

SEPTEMBER'S CONSTELLATION

Look almost overhead for five stars in a distinctive W-shape. To the Greeks, this constellation represented Queen **Cassiopeia** of Ethiopia, who ruled with her husband King **Cepheus**. The hero **Perseus** rescued their daughter **Andromeda** from a ravaging sea monster (**Cetus**), and these characters are now all immortalised in the heavens.

Unusually, the central star in Cassiopeia is known internationally by its Chinese name, **Tsih** (the Whip). This unstable star – 34,000 times brighter than the Sun – is spinning around at breakneck pace, flinging out streams of gas.

Cassiopeia has hosted several supernovae, including an exploding star seen by Danish astronomer Tycho Brahe in 1572. The fireball from a later supernova is now the most prominent radio source in the sky, Cassiopeia A.

OBSERVING TIP

When you first go out to observe, you may be disappointed at how few stars you can see in the sky. But wait for 20 minutes, and you'll be amazed at how your night vision improves. One reason for this 'dark adaptation' is that the pupil of your eye grows larger. More importantly, in dark conditions the retina of your eye builds up bigger reserves of rhodopsin, the chemical that responds to light.

SEPTEMBER'S OBJECT

At first glance, the star **Delta Cephei** – in the constellation representing King Cepheus – seems pretty ordinary. At magnitude +4, this yellow star is visible to the naked eye, though it's not prominent. A telescope reveals a companion star. But this star holds the key to measuring the size of the Universe.

Check Delta Cephei carefully, and you'll see its brightness changes regularly, from +3.5 to +4.4, every 5 days 9 hours. It's a result of the star swelling and shrinking, from 32 to 35 times the Sun's diameter.

The American astronomer Henrietta Leavitt found that stars like this – cepheids – show a link between their period of variation and their intrinsic luminosity. By observing the star's period and its apparent brightness, astronomers can work out a cepheid's distance. With the Hubble Space Telescope, astronomers have now measured cepheids in the remote galaxy NGC 5468, some 130 million light years away.

SEPTEMBER'S TOPIC: DOUBLE STARS

The Sun is an exception in living a solitary life. Over half the stars you see in the sky are paired up, as double stars. In some cases, it's just that a distant star happens to lie almost behind a nearer one. To the naked eye, **Algedi** in **Capricornus** looks double, but the fainter star lies nine times further off than the brighter component.

The most famous celestial double act lies in the **Plough** in **Ursa Major**. **Mizar** (the brighter star) and its fainter companion **Alcor** have often been named 'the horse and rider'. These two stars are travelling together through space.

Most double stars, though, are orbiting each other – astronomers call them 'binary stars'. Check out the star marking the head of **Cygnus** (the Swan), **Albireo**, with a small telescope and you're in for a treat: a golden star with a sapphire companion. In the case of Spica (in Virgo) the two stars are so close that no telescope can separate them: the only clue it's double comes from examining Spica's light in detail, and seeing evidence of two stars moving alternately towards and away from us.

And binary stars are just the first step. The **Double-Double** in **Lyra** is exactly what its name proclaims: two pairs of stars in orbit around each other. Castor, one of the twin stars of Gemini, consists of three pairs of binary stars – making it a sextuple star.

SEPTEMBER'S PICTURE

Located in the obscure constellation of **Vulpecula** (the Fox) – next to the head of diminutive **Sagitta** (the Arrow) – the **Dumbbell Nebula** is a huge hit with stargazers. At magnitude +7.5, it's visible through binoculars, and is a great target to capture on camera, as exemplified in Andy Weller's atmospheric image.

For all its beauty, the Dumbbell is a ghostly wraith. This 'planetary nebula' is

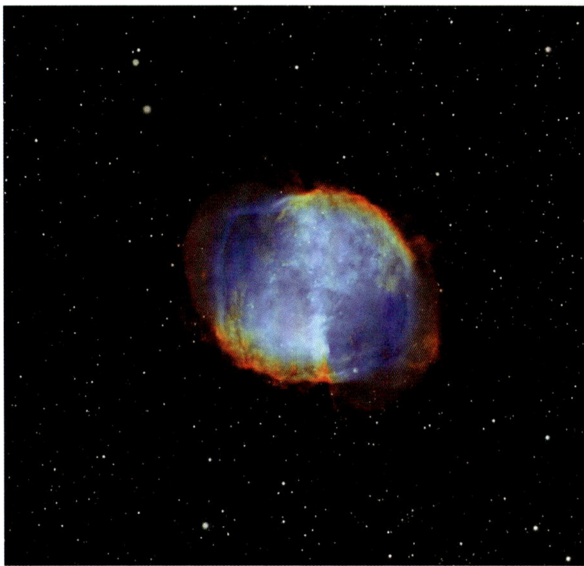

Andy Weller captured the Dumbbell Nebula with a Celestron C11 XLT 279-mm Schmidt–Cassegrain telescope and a ZWO ASI294MC Pro camera equipped with an Optolong L-eXtreme filter that passes only red hydrogen alpha light and green oxygen light ([OIII]). He took 90 × 120-second exposures, for a total exposure time of 3 hours, and processed them in PixInsight.

the remains of a star that has died. The term was first coined by planet-discoverer William Herschel, who noted the resemblance of these circular nebulae to planets; it was his son John who nicknamed the Dumbbell, because 'it resembled a double-headed shot'.

This nebula lies roughly 1360 light years away, and is an estimated 2.5 light years across. It was created when an ageing star ran out of hydrogen fuel in its core, and puffed off its enveloping atmosphere into space. From the expansion rate of the nebula, astronomers estimate that the fatal act took place some 10,000 years ago.

SEPTEMBER'S CALENDAR

SUNDAY	MONDAY	TUESDAY	WEDNESDAY	THURSDAY	FRIDAY	SATURDAY
		1 Venus near Spica	2 Moon near Pleiades	3 Moon near Pleiades	4 8.51 am Last Quarter Moon	5
6	7 Moon near Mars (am)	8 Moon occults Praesepe near Jupiter	9 Moon near Jupiter (am); maximum of 45P-ids	10	11 4.27 am New Moon	12
13	14 Moon near Venus	15	16	17 Moon near Antares	18 9.44 pm First Quarter Moon	19
20	21	22 Venus at maximum brightness	23 Autumn Equinox	24	25 Neptune opposition	26 5.49 pm Full Moon near Saturn
27 Moon near Saturn	28	29	30 Moon near Pleiades			

SPECIAL EVENTS

- **8 September, 5 am:** the crescent Moon occults the outer regions of the star cluster Praesepe, with Jupiter below (Chart 9a).
- **9 September, before dawn:** Jupiter lies above the crescent Moon.
- **9 September:** you may catch a few meteors streaming outwards from the border of Pisces and Aquarius in the **45P-id meteor shower**, as the Earth crosses a smattering of dust particles shed by comet 45P/Honda–Mrkos–Pajdušáková in 1927.
- **14 September:** Venus lies close to the crescent Moon just after sunset (Chart 9b).
- **22 September:** Venus reaches its greatest brilliance as an Evening Star this year (see Planet Watch).
- **23 September, 1.05 am:** nights become longer than days as the Sun moves south of the Equator at the Autumn Equinox.
- **25 September:** Neptune is nearest to the Earth at 4470 million km and opposite to the Sun (see Planet Watch).

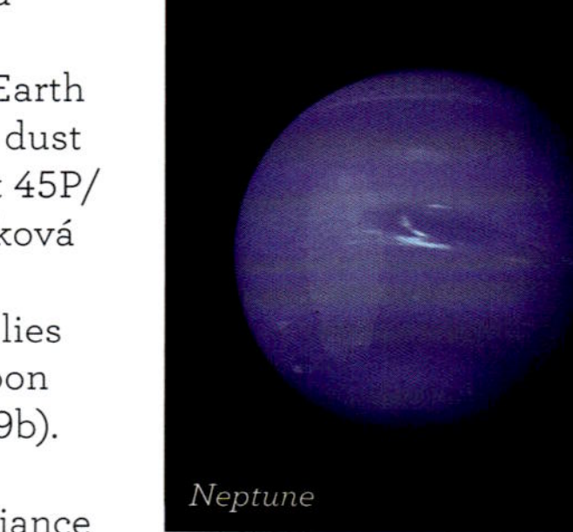

Neptune

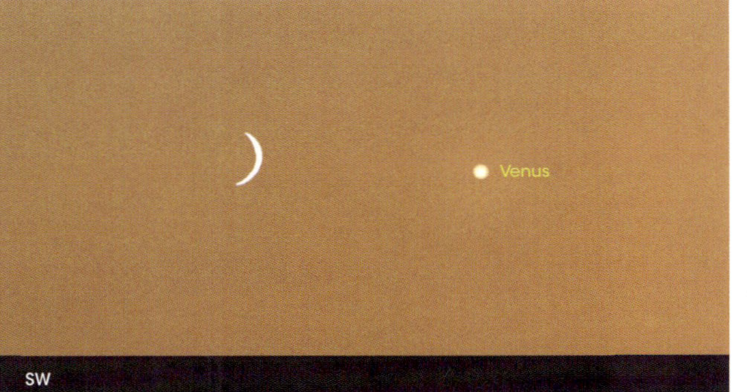

9a 8 September, 5 am. The Moon occults Praesepe, near Jupiter.

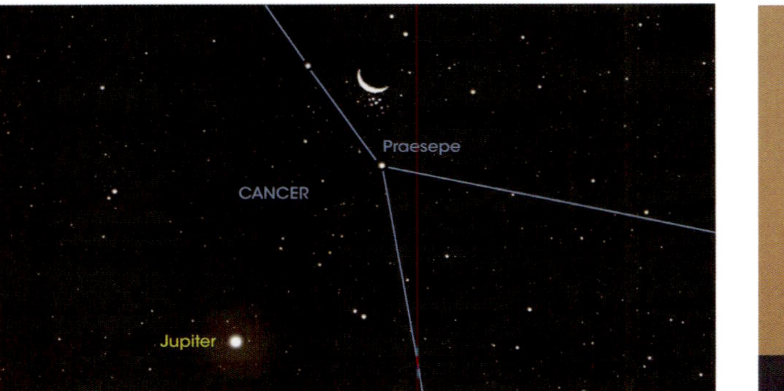

9b 14 September, 7.30 pm. The Moon near Venus.

- Starting the month just below Spica and setting around 8 pm, **Venus** drops rapidly into the dusk twilight. The crescent Moon lies to its left on 14 September (Chart 9b). Through a small telescope (or good binoculars), you can see the planet's crescent growing thinner as Venus starts to swing between the Earth and the Sun. The Evening Star reaches its maximum brilliance (magnitude –4.8) on 22 September, just as it disappears into the Sun's glare.

- **Saturn** (magnitude +0.4) rises about 8 pm in Cetus (not one of the conventional zodiacal signs). The Moon is near the ringworld on 26 and 27 September.

- Ten degrees to the right of Saturn, in Pisces, **Neptune** reaches opposition and its closest point to the Earth on 25 September (see Special Events), when it is visible all night long. Even at its brightest, though, the most distant planet is too faint to be seen without optical aid, peaking at magnitude +7.7.

- **Uranus** lies in Taurus, on the borderline of naked-eye visibility at magnitude +5.7, and rises around 9.30 pm.

- Rising about 1 am, **Mars** travels rapidly through Gemini during September: at magnitude +1.2 it's about the same brightness as the constellation's leading stars, Castor and Pollux. On the morning of 7 September, the Moon lies to the left of Mars with Castor and Pollux above. The Red Planet is directly in line with the twin stars in the early hours of 25 September.

- **Jupiter** is now brilliant in the morning sky, rising about 3 am and blazing at magnitude –1.8 in Cancer. The Moon is nearby on 8 September (Chart 9a) and 9 September.

- **Mercury** is too close to the Sun to be visible this month.

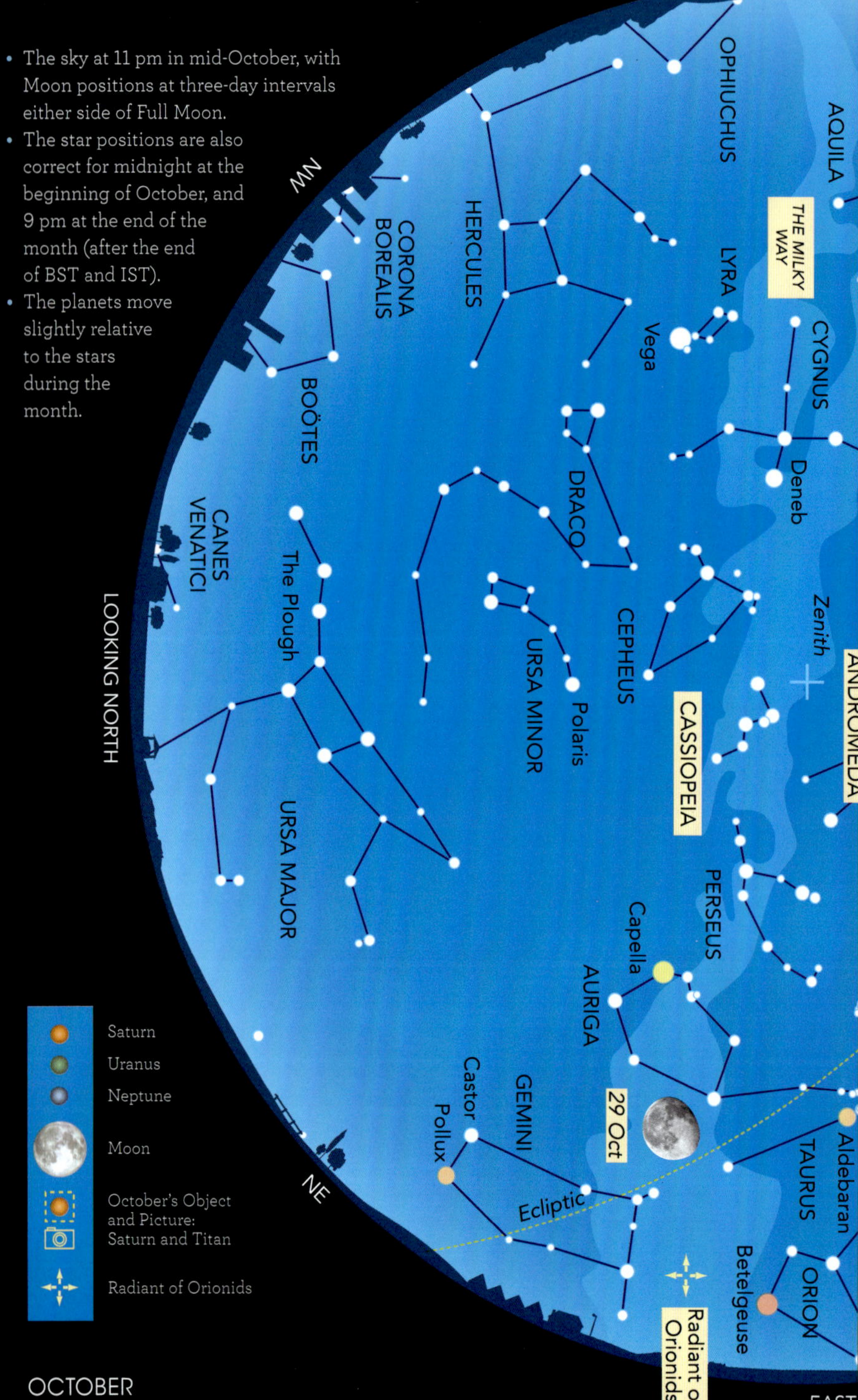

- The sky at 11 pm in mid-October, with Moon positions at three-day intervals either side of Full Moon.
- The star positions are also correct for midnight at the beginning of October, and 9 pm at the end of the month (after the end of BST and IST).
- The planets move slightly relative to the stars during the month.

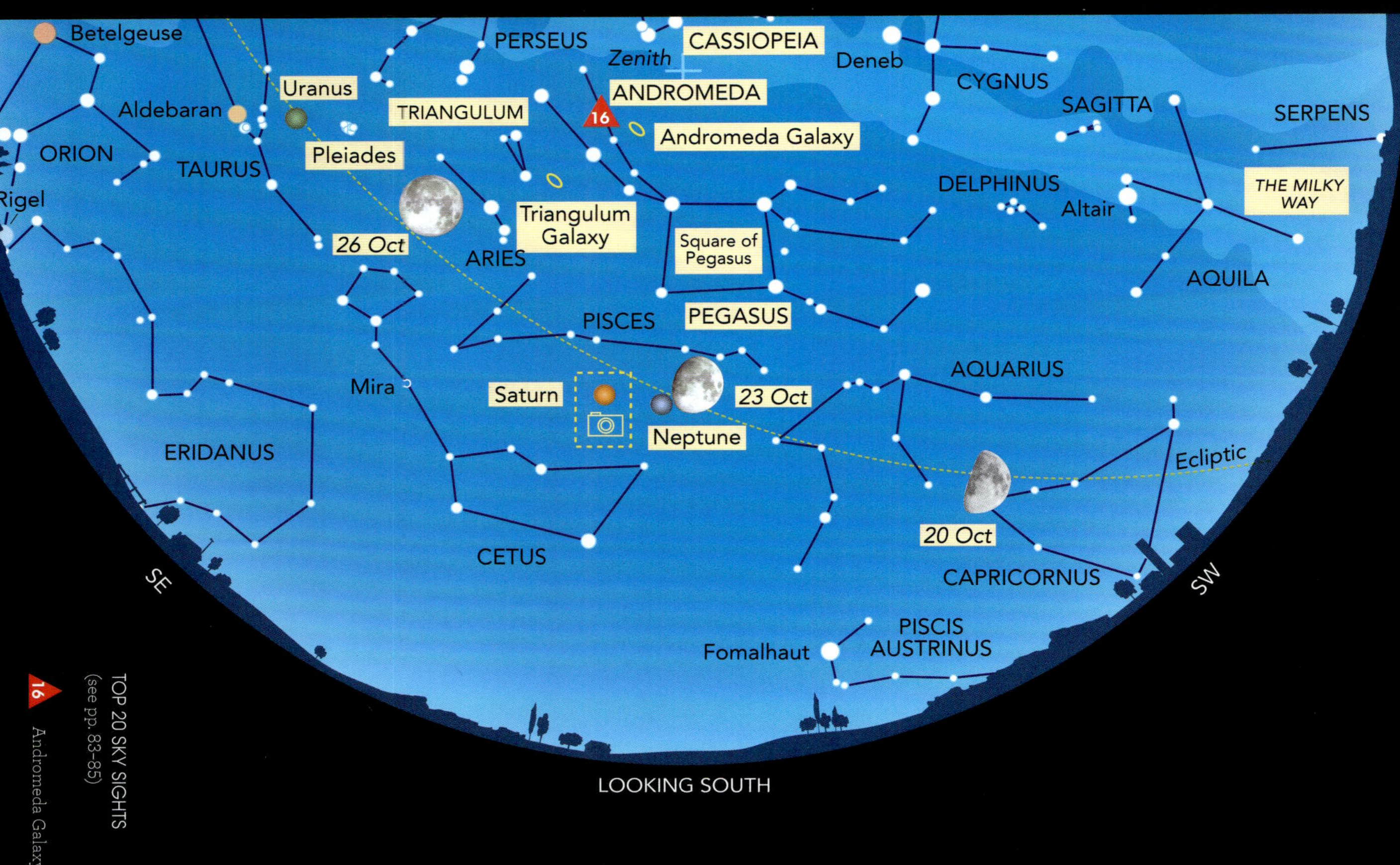

EAST
WEST
Betelgeuse
PERSEUS
CASSIOPEIA
Zenith
Deneb
CYGNUS
Uranus
ANDROMEDA
SAGITTA
Aldebaran
TRIANGULUM
16
SERPENS
ORION
Andromeda Galaxy
TAURUS
Pleiades
DELPHINUS
Rigel
Altair
THE MILKY WAY
Triangulum Galaxy
Square of Pegasus
26 Oct
ARIES
AQUILA
PISCES
PEGASUS
AQUARIUS
Mira
Saturn
23 Oct
Neptune
ERIDANUS
Ecliptic
20 Oct
CETUS
CAPRICORNUS
SE
SW
PISCIS AUSTRINUS
Fomalhaut
16
TOP 20 SKY SIGHTS
(see pp. 83–85)
Andromeda Galaxy
LOOKING SOUTH
OCTOBER
63
OCTOBER

Saturn is closest to the Earth this month: grab a telescope to check out its renowned rings. The autumn constellations are pretty drab: neither the **Square of Pegasus** nor **Andromeda** are guaranteed to thrill. But look to the east, where the brilliant lights of winter are starting to appear, spearheaded by the beautiful star cluster of the **Pleiades**.

OCTOBER'S CONSTELLATION

It may be small, but it's venerable. **Triangulum** (the Triangle) dates back three millennia to ancient Babylonian astronomers, who envisaged its stars as the sharp end of a plough that also included stars in **Andromeda** and **Cassiopeia**. The geometrical Greeks saw a triangle here, sometimes described as representing the shape of Sicily.

The constellation's most noteworthy object is the **Triangulum Galaxy** (catalogued as M33). Just visible to the naked eye on the darkest nights, the galaxy is an easy sight in binoculars. The Triangulum Galaxy is an untidy spiral, the third-largest member of our Local Group of galaxies (after the **Andromeda Galaxy** and the **Milky Way**), and lies almost 3 million light years away.

OCTOBER'S OBJECT

Glorious ringworld **Saturn** takes centre stage in the heavens this month, reaching opposition (highest in the south at midnight) on 4 October. After Jupiter, Saturn is the second-largest planet. But its density is so low that – were you to plop it in an ocean – Saturn would float. The planet spins around in only 10 hours 33 minutes, whipping up winds that roar around Saturn at 1800 kilometres per hour.

Through a small telescope you can spot the dazzling icy rings that engirdle the planet, and also Saturn's largest moon, Titan (see Picture). Saturn boasts a family of 274 moons – the highest total for any planet in the Solar System – and they are a fascinating bunch. Iapetus is brilliant white on one side and black on the other, for instance, while Enceladus is spewing salty water into space and probably has a liquid ocean under its icy surface.

OCTOBER'S TOPIC: JOCELYN BELL BURNELL

Young research student Jocelyn Bell (later Bell Burnell, 1943–) couldn't believe the paper chart that was spooling out of her radio telescope. It showed pulses of radio waves from space,

OBSERVING TIP

The Andromeda Galaxy is often described as the furthest object 'easily visible to the unaided eye'. It can be elusive, though – especially if you are suffering from light pollution. The trick is to memorise Andromeda's pattern of stars, and then to look slightly to the side of where you expect the galaxy to be. This technique – called 'averted vision' – causes the image to fall on the outer region of your retina, which is more sensitive to light than the central region that's evolved to discern fine details. Averted vision is also crucial when you want to observe the faintest nebulae or galaxies through a telescope.

repeating so precisely every 1.33 seconds that they looked like an artificial signal. Half-jokingly, she labelled this radio source LGM-1, for Little-Green-Men-1.

The reality was almost as strange. The 'pulsar' signals were coming from the shrunken corpse of a star, as heavy as the Sun but only the size of a city. This neutron star was emitting beams of radiation that swept past the Earth regularly as the star spun round, like the rotating beams from a lighthouse, and hence produced pulses in the radio telescope.

Born in Northern Ireland, Bell Burnell went to university in Glasgow and then Cambridge, where she built the radio telescope that found pulsars. The discovery revolutionised our knowledge of star death, paving the way for the even more extraordinary black holes. Bell Burnell has gone on to a stellar astronomical career investigating other exotic astronomical objects, while also promoting women and other minorities in science. A bank in Ireland has commemorated Bell Burnell by putting her image on their £50 note.

OCTOBER'S PICTURE

Two unique Solar System objects appear together in this image: the brilliant icy rings of Saturn and Titan, the only moon with a substantial

atmosphere. Damian Peach captured them when they appeared close together, but in reality Titan is over a million kilometres behind Saturn in this image.

Saturn's rings are composed of over a million trillion chunks of ice orbiting the planet as 'mini moons', with sizes ranging from smaller than a pea to the dimension of a house. The shiny freshness of the rings suggests that they formed less than 100 million years ago when a small icy moon was shattered, possibly when it was hit by a comet.

Bigger than the planet Mercury, Titan is cloaked in an atmosphere of nitrogen, with a pressure 50 per cent greater than that of Earth's air. But there the similarities end. Titan is a moon in deep-freeze: it struggles in temperatures of −180°C, with seas and lakes of liquid ethane and methane. In 2005, the Huygens probe penetrated Titan's murky orange clouds and landed on the moon's surface – a precursor to a future helicopter mission called Dragonfly that will explore Titan in intimate detail.

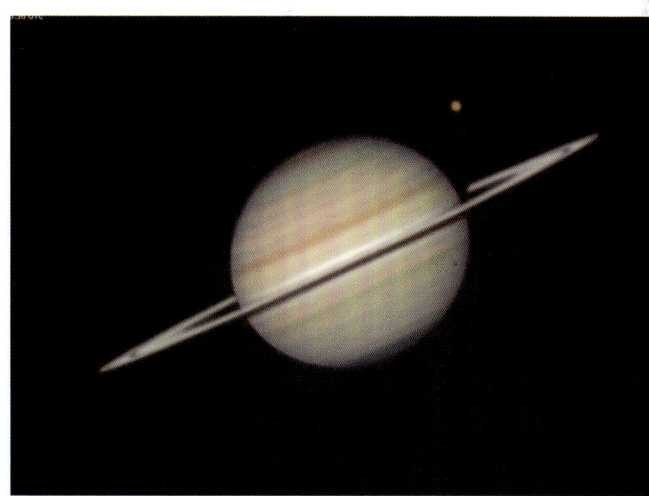

Titan was swinging behind Saturn on 24 July 2024, when Damian Peach captured this image with a Celestron 356-mm Schmidt–Cassegrain telescope and QHY5III200M camera equipped with red, green and blue (RGB) filters. Damian stacked thousands of frames from a high-speed video to create the final image.

OCTOBER'S CALENDAR

SUNDAY	MONDAY	TUESDAY	WEDNESDAY	THURSDAY	FRIDAY	SATURDAY
				1	2	3 2.25 pm Last Quarter Moon
4 Saturn opposition	5 Moon very near Mars (am)	6 Moon near Jupiter (am)	7 Moon near Regulus (am)	8	9	10 4.50 pm New Moon
11 Mars in front of Praesepe (am)	12 Mercury E elongation	13	14 Moon near Antares	15	16	17
18 5.12 pm First Quarter Moon	19	20	21 Orionids	22 Orionids (am)	23 Moon near Saturn	24 Moon near Saturn
25 BST and IST end	26 4.12 am Full Moon	27 Moon occults the Pleiades	28 Moon occults the Pleiades (am)	29	30	31

SPECIAL EVENTS

- **4 October:** Saturn lies opposite the Sun in the sky, and closest to Earth at 1262 million km.
- **5 October, 6 am:** the Moon skims over Mars, at separation of only 25 arcminutes (Chart 10a).
- **6 October, before dawn:** Jupiter lies just below the crescent Moon.
- **11 October, before dawn:** Mars appears to lie in the middle of Praesepe (see Planet Watch and Chart 10b).
- **Night of 21/22 October:** high-speed debris from Halley's Comet smashes into the Earth's atmosphere as the annual **Orionid meteor shower**, best seen in the early morning hours after the Moon has set.
- **23 October:** you'll find Saturn to the left of the Moon.
- **24 October:** the Moon lies above Saturn.
- **25 October, 2 am:** the end of British Summer Time and Irish Standard Time.
- **Night of 27/28 October, 11.30 pm–3 am:** the Moon occults the upper stars of the Pleiades.

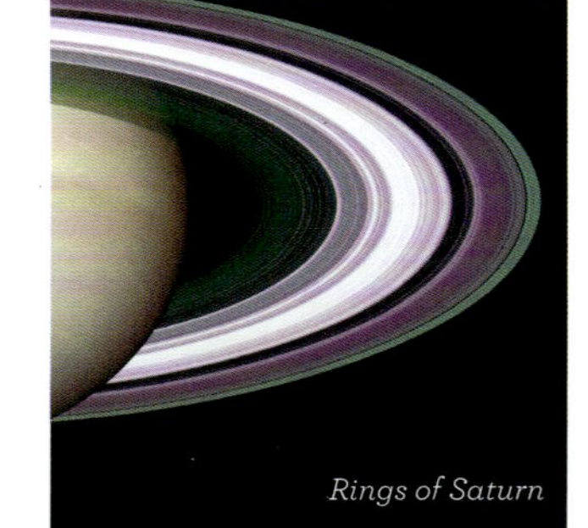

Rings of Saturn

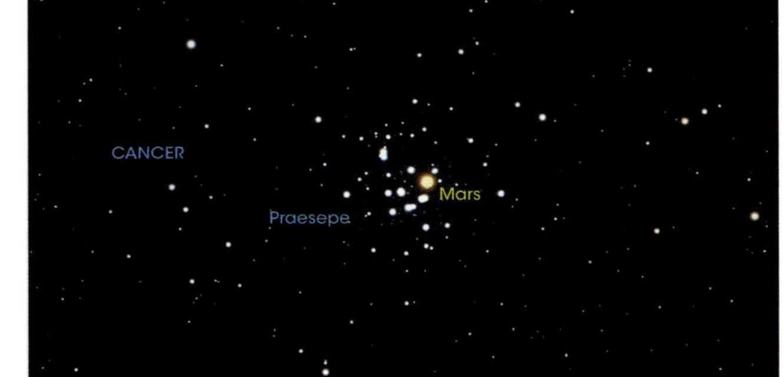

10a 5 October, 6 am. The Moon passes close to Mars.

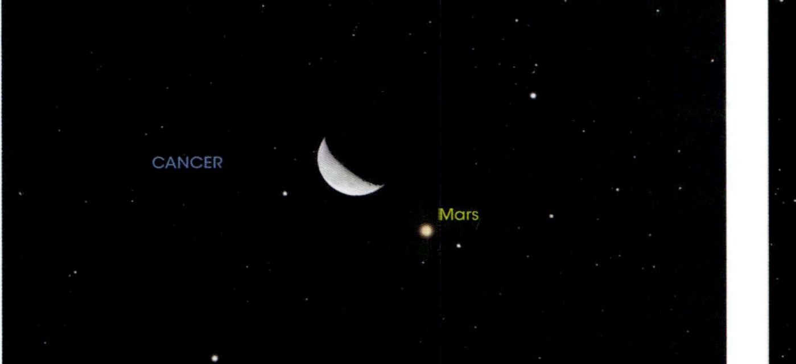

10b 11 October, 6 am. Mars lies in front of Praesepe (binocular view).

- **Saturn** is at opposition on 4 October, visible all night long in Cetus (just to the south of the zodiacal constellation Pisces). At its nearest to Earth (see Special Events), the planet reaches a maximum magnitude of +0.3. The Moon lies near Saturn on 23 and 24 October. Through a telescope you can make out the planet's famous rings and some of its larger satellites, including giant Titan which is swathed in dense orange clouds.
- Faint **Neptune**, in Pisces, sulks at a mere magnitude +7.7 and sets around 5.30 am.
- **Uranus** is seven times brighter, and on the border of naked-eye visibility at magnitude +5.6. The seventh planet lies in Taurus, and rises about 7.30 pm.
- Rising around 0.30 am, **Mars** (magnitude +1.0) is moving through Cancer this month. On the morning of 5 October, the Moon passes just above the Red Planet (see Special Events and Chart 10a). In the wee small hours of 11 October, Mars lies smack-bang in front of Praesepe: through binoculars or a low-power telescope, it's a ruby set in a field of diamonds (Chart 10b). Towards the end of the month, Mars is closing in on Jupiter.
- The giant of the Solar System, **Jupiter** rises about 2 am in Leo, resplendent at magnitude –1.9. The crescent Moon lies above Jupiter on the morning of 6 October. On the last few mornings of October, you'll find Jupiter at the centre of a ménage-à-trois, below Mars and above Regulus.
- **Mercury** and **Venus** are lost in the Sun's glare in October, even though Mercury reaches its greatest separation from the Sun on 12 October.

OCTOBER'S PLANET WATCH

- The sky at 10 pm in mid-November, with Moon positions at three-day intervals either side of Full Moon.
- The star positions are also correct for 11 pm at the beginning of November, and 9 pm at the end of the month.
- The planets move slightly relative to the stars during the month.

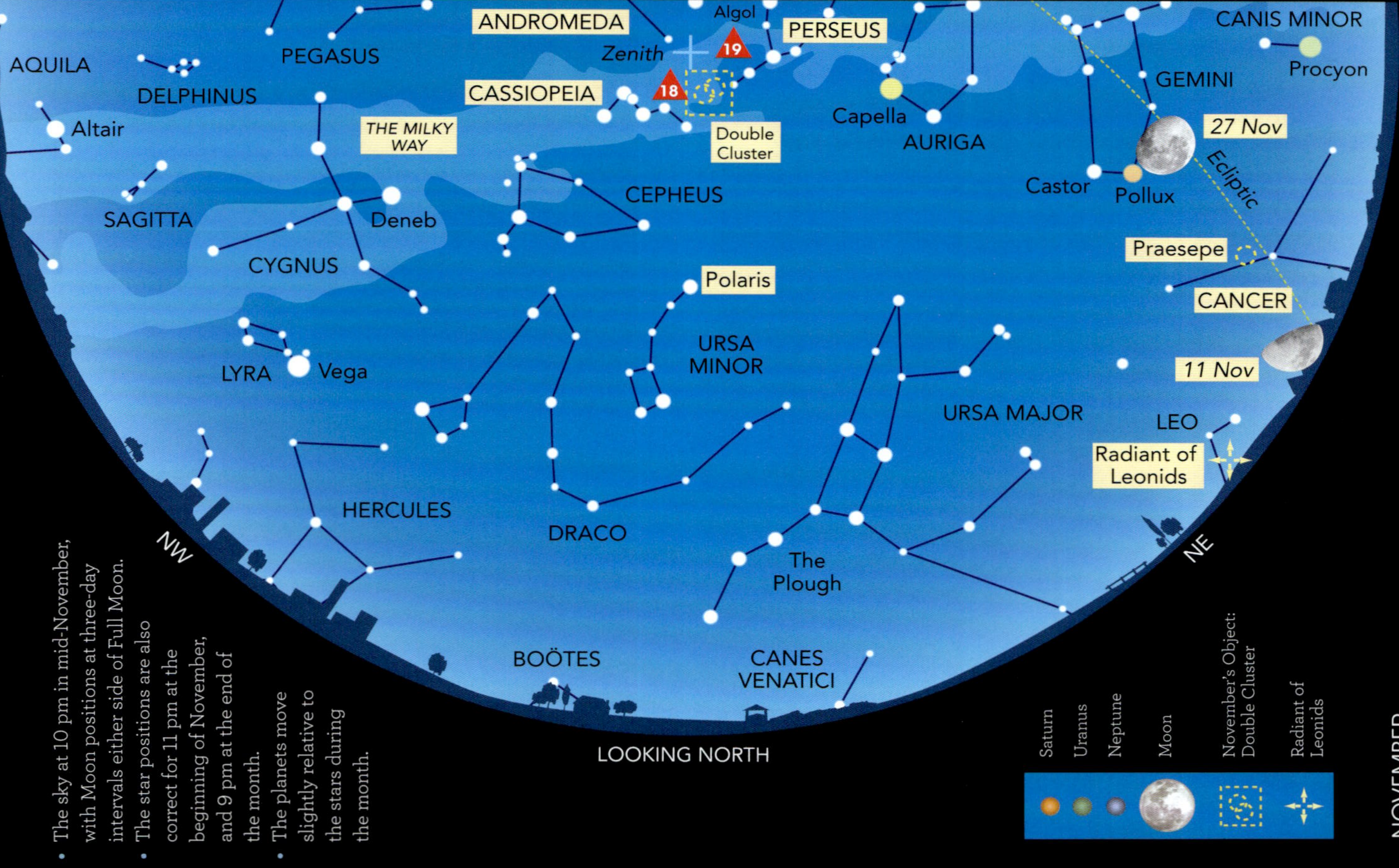

EAST
WEST
Procyon
CANIS MINOR
AURIGA
Capella
Double Cluster
18
Zenith
CASSIOPEIA
CYGNUS
DELPHINUS
AQUILA
PERSEUS
ANDROMEDA
Algol
19
24 Nov
THE MILKY WAY
TAURUS
Pleiades
TRIANGULUM
ARIES
Square of Pegasus
PEGASUS
Betelgeuse
Aldebaran
Uranus
Hyades
ORION
21 Nov
PISCES
AQUARIUS
Menkar
CAPRICORNUS
18 Nov
Ecliptic
Saturn
Neptune
Rigel
Mira
17
LEPUS
ERIDANUS
CETUS
Fomalhaut
SE
SW
LOOKING SOUTH

TOP 20 SKY SIGHTS
(see pp. 83–85)

17   Mira
18   Double Cluster
19   Algol

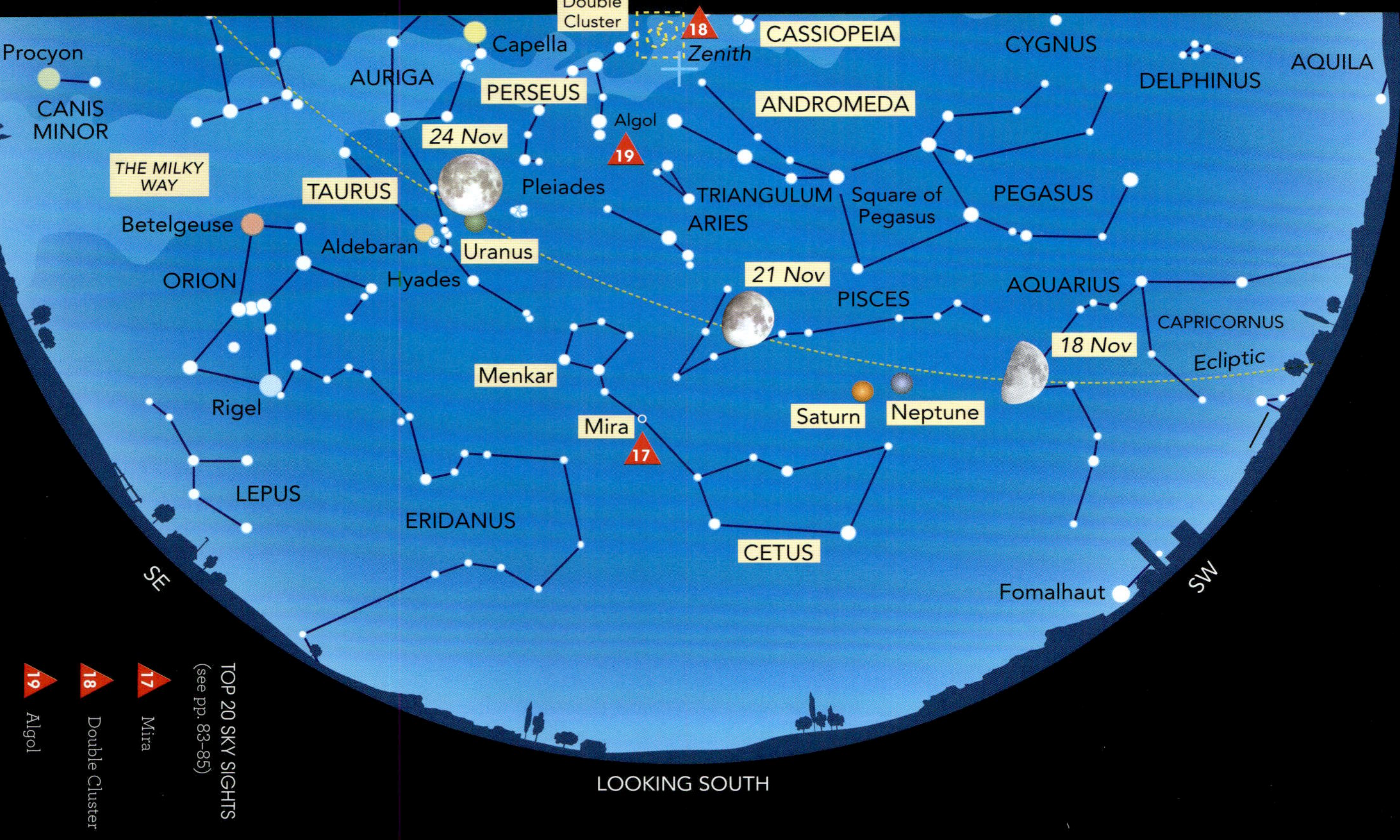

NOVEMBER

It's a great month for checking out the **Milky Way**, rearing overhead on these long November nights and providing an inside perspective on our own home Galaxy. Sweep along the glowing band with binoculars, and you'll find it studded with star clusters and nebulae.

NOVEMBER'S CONSTELLATION

In Greek mythology, **Cetus** (the Sea Monster) was a fearsome creature, sent to devour princess **Andromeda** when her mother **Cassiopeia** boasted she was more beautiful than the sea nymphs. Fortunately, the passing superhero **Perseus** slew the ravening beast.

While the large dull constellation hardly lives up to this billing, one highlight is **Menkar**, marking the monster's tail. Through a telescope you'll see it is a pretty orange-and-blue double, though these stars are at different distances.

Another highlight is **Mira** (the Wonderful). A distended red giant, some 400 times wider than the Sun, Mira shrinks and swells over a period of 11 months. At the moment, Mira is around minimum brightness (magnitude +10), hardly visible even in binoculars. But keep watching Cetus over the winter, as

Mira surges to its next maximum early in 2027, when it's likely to rival **Polaris** at magnitude +2.

NOVEMBER'S OBJECT

A pair of objects this month: the beautiful **Double Cluster** in Perseus, on the border with Cassiopeia. Visible to the unaided eye, these near-twin star clusters – each covering an area bigger than the Full Moon – are a gorgeous sight in binoculars. Each cluster is loaded with glorious young blue supergiant stars, with a sprinkling of red giants to add to their visual appeal.

The cluster h Persei (otherwise known as NGC 869) is 7460 light years away from us, and chi Persei (NGC 884) is slightly further away at 7640 light years. They form the core of the Perseus OB1 Association, an extensive group of bright, hot stars that were born together 14 million years ago.

NOVEMBER'S TOPIC: SETI

It's the biggest question in astronomy: is there anybody out there? Over 60 years ago, American astronomer Frank Drake kickstarted the Search for Extraterrestrial Intelligence (SETI) when he turned his radio telescope to the heavens in the hope of hearing an alien broadcast. He was met with a deafening silence.

Since then, astronomers have used ever more sensitive instruments to tune into the sky, including a purpose-built array of dishes in California, and the

---

OBSERVING TIP

With Christmas on the way, you may be thinking of buying a telescope as a present for a budding stargazer. Beware! Unscrupulous websites and mail-order catalogues often advertise small telescopes that boast huge magnifications, but all they do is show you a large blurry image. To see planets and stars sharply, you need to use a magnification no more than twice the diameter of the lens or mirror in millimetres. So if you see an advertisement for a 75-mm telescope, beware of any claims for a magnification greater than 150 times.

---

world's largest radio telescope, the Five-hundred-meter Aperture Spherical Telescope (FAST) in China. The Breakthrough Listen project is checking out a million stars in our Galaxy, plus a hundred of the nearest galaxies, for alien radio signals. In 2024, astronomers started to use artificial intelligence to analyse the mountains of data in unprecedented detail.

Other researchers have tried to pick up flashes of laser light from extraterrestrial civilisations. Scientists have also checked whether unusual-looking stars may be surrounded by giant structures built by aliens. And there's a search for extraterrestrial spacecraft speeding through the Solar System.

So far, all to no avail. Maybe the aliens are using technology far beyond our comprehension? Or they may be intentionally minimising contact with us, to observe our primitive ways, as we monitor animals in the wild. Or, most sobering of all, perhaps we are the only intelligent life form in our Galaxy.

NOVEMBER'S PICTURE

When someone tells me the Moon looks really weird tonight, I can almost guarantee that – if a lunar eclipse isn't due – what they are seeing is a lunar halo. It's a glowing ring around the Moon, with a radius roughly equal to the length of your outstretched fingers, and beautifully caught here by Mary McIntyre.

Although the Moon's light is ultimately responsible, the halo is not an astronomical phenomenon. It's caused by a cloud of tiny ice crystals high in the Earth's atmosphere. As moonlight passes through each hexagonal crystal, refraction bends its path through an angle of 22 degrees. Ice crystals that happen to lie this distance from the Moon in the sky refract moonlight in our direction, and so we see a bright ring of light with a radius of 22 degrees around the Moon itself. Different wavelengths are refracted by slightly different amounts, so the inner edge of the halo is reddish and the outer edge more bluish.

Icy cirrus clouds can form ahead of a meteorological front, so often a lunar halo portends bad weather – hence its alternative name, a 'storm ring'.

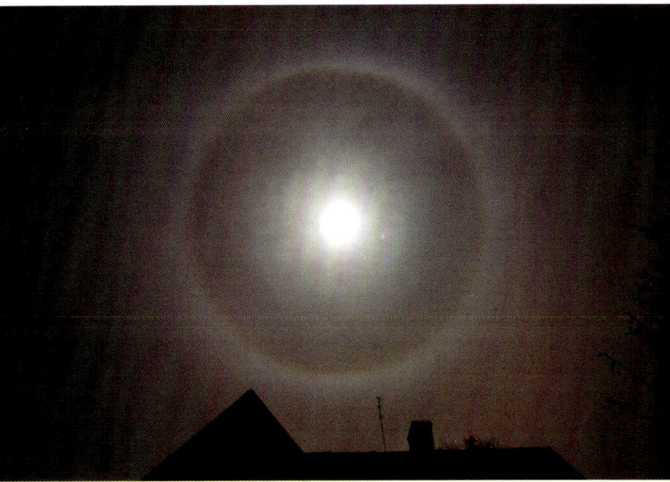

On the night of 23 November 2023, 'we had a lot of different atmospheric optics on display,' recalls Mary McIntyre, who photographed this lunar halo over her home in Oxfordshire with a Canon 1100D camera fitted with a Canon 10-mm lens. It was a single 4-second exposure at ISO 800 and aperture f/4.5, processed in Lightroom and FastStone Image Viewer.

NOVEMBER'S CALENDAR

SUNDAY	MONDAY	TUESDAY	WEDNESDAY	THURSDAY	FRIDAY	SATURDAY
1 8.28 pm Last Quarter Moon	**2** Moon near Mars (am)	**3** Moon near Jupiter and Regulus (am)	**4**	**5**	**6**	**7** Moon near Venus and Spica (am)
8	**9** 7.02 am New Moon	**10**	**11**	**12**	**13**	**14**
15	**16** Mars near Jupiter	**17** 11.48 am First Quarter Moon; Leonids	**18** Leonids (am)	**19**	**20** Mercury W elongation; Moon near Saturn	**21**
22	**23**	**24** 2.53 pm Full Moon; supermoon	**25** Uranus opposition	**26** Uranus nearest to Earth; Mars near Regulus	**27** Venus at maximum brightness	**28** Moon occults Praesepe
29 Moon near Jupiter, Mars and Regulus	**30** Moon near Jupiter, Mars and Regulus					

SPECIAL EVENTS

- **2 November, am:** the Moon lies above Mars, Jupiter and Regulus (Chart 11a).
- **3 November, am:** the Moon is close to Jupiter, Regulus and Mars (Chart 11a).
- **7 November, before dawn:** the crescent Moon teams up with Venus, next to Spica (Chart 11b).
- **16 November:** Mars passes a degree to the left of Jupiter.
- **Night of 17/18 November:** the maximum of the **Leonid meteor shower,** fiery fragments of Comet Tempel–Tuttle, best seen after moonset at 11.30 pm.
- **20 November, before dawn:** Mercury reaches its greatest morning separation from the Sun.
- **24 November:** the year's second supermoon (see December's Special Events).
- **25 November:** Uranus is opposite to the Sun.
- **26 November:** Uranus is closest to the Earth at 2,759 million km; Mars passes 1.5 degrees to the left of Regulus.
- **27 November:** Venus at its greatest brilliance as a Morning Star.
- **28 November, 11 pm:** the Moon passes in front of Praesepe.
- **29 November:** the Moon lies close to Jupiter, Mars and Regulus.
- **30 November:** the Moon is just below Jupiter, Mars and Regulus.

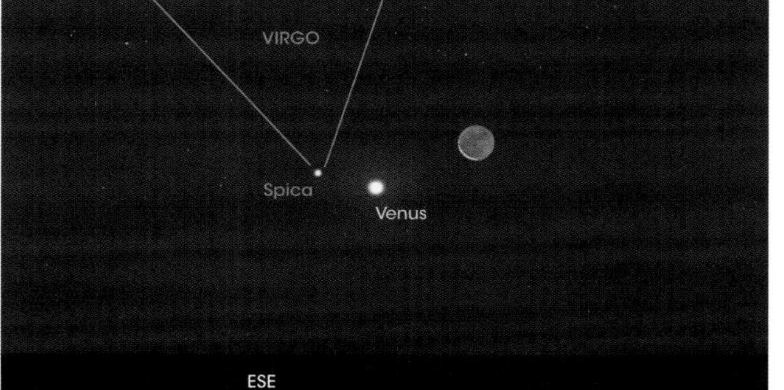

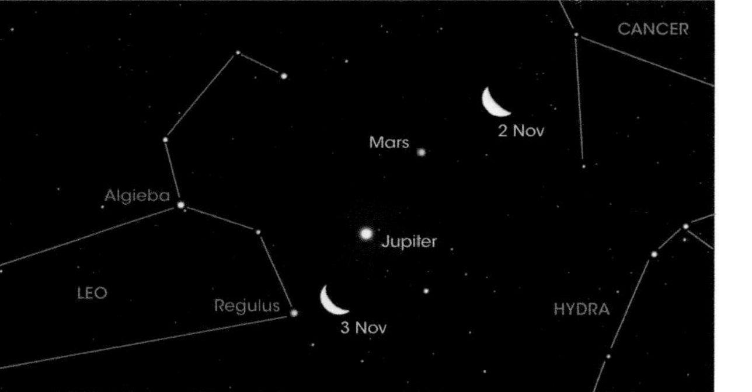

11a 2–3 November, 5 am. The Moon with Mars, Jupiter and Regulus.

11b 7 November, 6 am. The Moon near Venus and Spica.

- The only planet easily visible in the evening sky is **Saturn** (magnitude +0.5), lying in Cetus and setting around 3 am. The Moon is near Saturn on 20 November.
- In the adjacent constellation Pisces, **Neptune** is only visible with binoculars or a telescope as it glows dimly at magnitude +7.7. The most distant planet sets about 2.30 am.
- **Uranus** is closest to the Earth on 26 November, the day after it's opposite to the Sun (see Special Events). Above the horizon all night long, Uranus lies in Taurus between Aldebaran and the Pleiades. Reaching its maximum brightness of magnitude +5.6 this month, Uranus is visible to the naked eye under good conditions, especially as it's currently high in the sky.
- **Jupiter** and **Mars** are both in Leo, near to Regulus, and rising about 11 pm. Jupiter, at magnitude –2.0, is ten times brighter than the Red Planet (magnitude +0.6). At the start of the month, Mars is higher in the sky, but it's heading rapidly downwards and passes to the left of Jupiter on 16 November. The Moon lies nearby on the morning of 2 and 3 November (Chart 11a), and on the evenings of 29 and 30 November.
- **Venus** roars into view as the Morning Star this month, reaching its maximum brilliance (magnitude –4.9) on 27 November, when it's rising as early as 4 am. The Moon forms a lovely duo with Venus on the morning of 7 November next to Spica (Chart 11b). Watch Venus then loop around Spica throughout the rest of the month.
- During the second half of November, **Mercury** puts on its best morning appearance of the year, to the lower left of Venus. At its greatest elongation on 20 November, it shines at magnitude –0.5, rising about 5.30 am.

NOVEMBER'S PLANET WATCH

- The sky at 10 pm in mid-December, with Moon positions at three-day intervals either side of Full Moon.
- The star positions are also correct for 11 pm at the beginning of December, and 9 pm at the end of the month.
- The planets move slightly relative to the stars during the month.

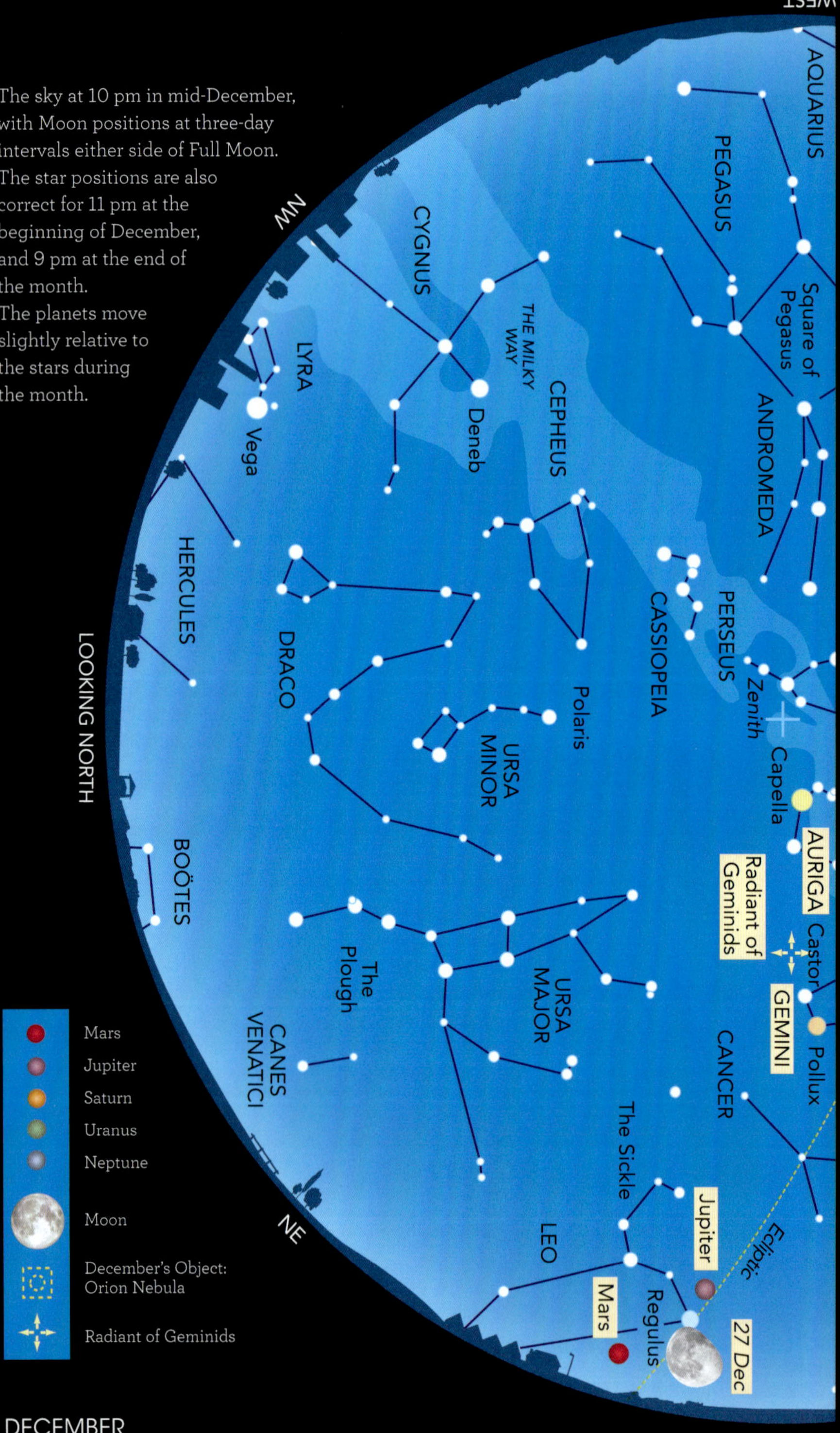

WEST

AQUARIUS

PEGASUS

Square of Pegasus

ANDROMEDA

NW

CYGNUS

THE MILKY WAY

LYRA

Vega

Deneb

CEPHEUS

PERSEUS

Zenith

Capella

AURIGA

Castor

Pollux

GEMINI

Radiant of Geminids

CANCER

HERCULES

DRACO

CASSIOPEIA

Polaris

URSA MINOR

LOOKING NORTH

BOÖTES

The Plough

URSA MAJOR

CANES VENATICI

LEO

The Sickle

Jupiter

Regulus

Mars

27 Dec

Ecliptic

NE

EAST

Mars
Jupiter
Saturn
Uranus
Neptune

Moon

December's Object:
Orion Nebula

Radiant of Geminids

74 DECEMBER

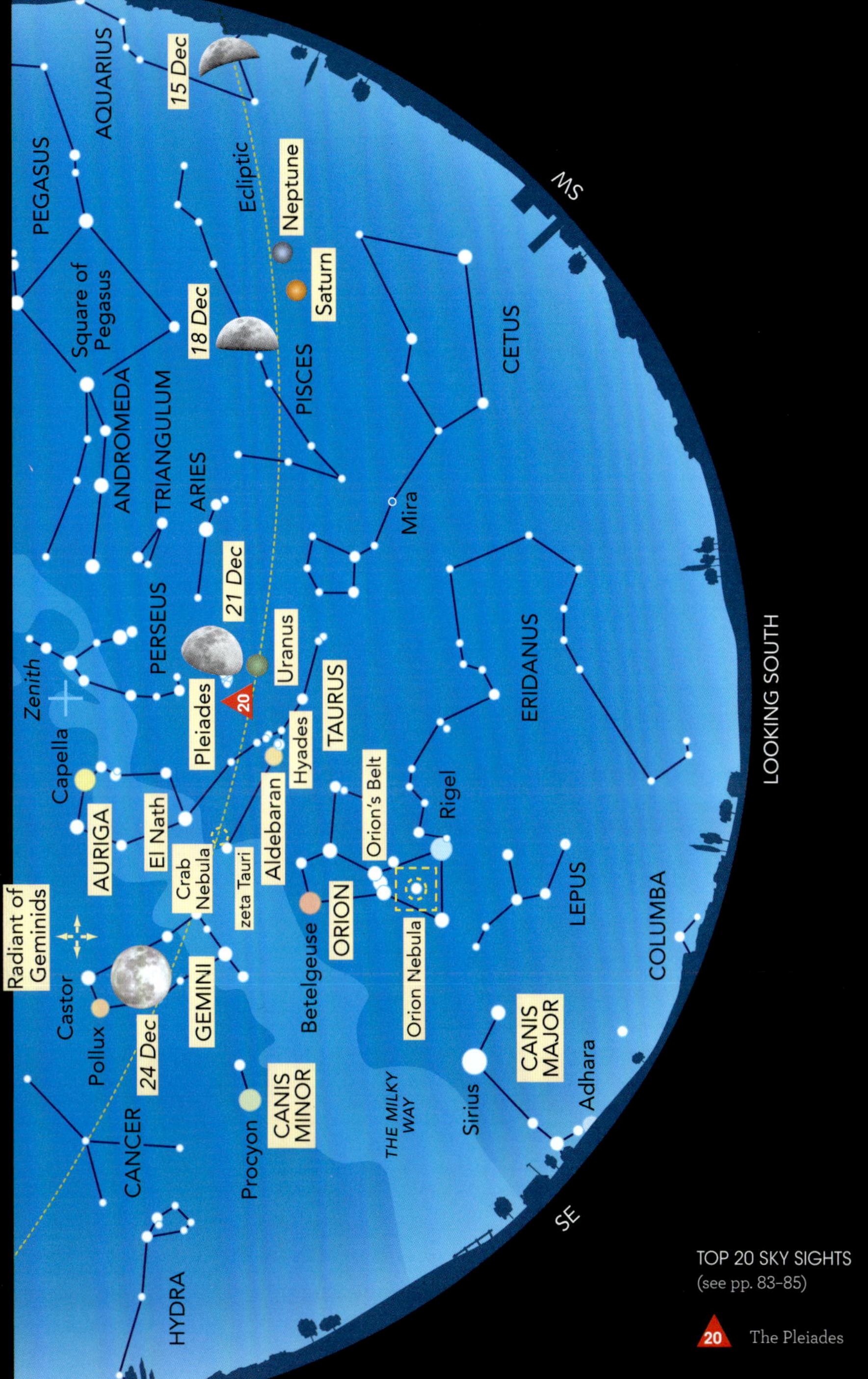

DECEMBER
WEST
PEGASUS
AQUARIUS
15 Dec
Square of Pegasus
ANDROMEDA
PERSEUS
TRIANGULUM
ARIES
18 Dec
Ecliptic
Neptune
Saturn
PISCES
CETUS
Mira
21 Dec
Pleiades
Uranus
TAURUS
ERIDANUS
Zenith
Capella
AURIGA
El Nath
Crab Nebula
zeta Tauri
Hyades
Aldebaran
Orion's Belt
Rigel
Radiant of Geminids
Castor
Pollux
24 Dec
GEMINI
Betelgeuse
ORION
Orion Nebula
LEPUS
COLUMBA
CANCER
CANIS MINOR
Procyon
THE MILKY WAY
Sirius
CANIS MAJOR
Adhara
HYDRA
EAST
SE
LOOKING SOUTH
SW
TOP 20 SKY SIGHTS
(see pp. 83–85)
20 The Pleiades

Look out for a brilliant meteor shower, plus the brightest Full Moon of the year. The stars, too, are putting on a celebratory show, featuring **Orion** with his hunting dogs **Canis Major** and **Canis Minor**, co-starring **Taurus** (the Bull), the hero twins of **Gemini** and the charioteer **Auriga**.

DECEMBER'S CONSTELLATION

Taurus is dominated by **Aldebaran**, the baleful eye of the celestial bovine, shining at magnitude +0.9. Although its hue is a distinct orange, Aldebaran is classified as a 'red giant' star, a bloated and more elderly version of our Sun.

The 'head' of the Bull is outlined by the **Hyades** star cluster, 153 light years away. Although Aldebaran looks to be part of this star grouping, it lies at less than half that distance. Even further off (440 light years) is the more spectacular **Pleiades** (or Seven Sisters) star cluster, packed with very young and brilliant stars.

Taurus has two 'horns': the star **El Nath** (Arabic for 'the butting one') to the north, and **zeta Tauri**, whose Babylonian name Shurnarkabti-sha-shutu – meaning 'star in the bull towards the south' – is thankfully not used very

much! Nearby, Chinese astronomers witnessed a brilliant supernova in 1054. The remains of this exploding star are visible today through a medium-sized telescope, as the still-glowing **Crab Nebula**.

DECEMBER'S OBJECT

The 17th-century Dutch astronomer Christiaan Huygens described the stars of Orion's 'sword' (hanging below **Orion's Belt**) as shining 'through a nebula so that the space around them seemed brighter than the rest of the heavens… the effect being that of an opening in the sky through which a brighter region was visible'.

Huygens was prescient. The **Orion Nebula** is a glowing region of gas in the midst of a huge black cloud, some 1340 light years away. The nebula itself – easily visible to the unaided eye under dark skies – is 25 light years across and is illuminated by a brilliant young star within, called Theta-1 Orionis C, which is 200,000 times brighter than the Sun and only a couple of million years old.

The surrounding dark cloud, the Orion-A Molecular Cloud, is a vast region of starbirth, the nearest location to Earth where heavyweight stars are being born. The Orion 'maternity ward' contains hundreds of fledgling stars, which have just hatched out of immense dark clouds of dust and gas; there's enough raw material here to make another 100,000 stars.

OBSERVING TIP

Hold a meteor party to check out the Geminid meteor shower on the night of 13/14 or 14/15 December. You don't need any optical equipment: the ideal viewing equipment is your unaided eye, plus a warm sleeping bag and a lounger. Everyone should look in different directions, so you can cover the whole sky. Shout out 'meteor!' when you see a shooting star. One of your party can record the observations, using a watch, notepad and red torch.

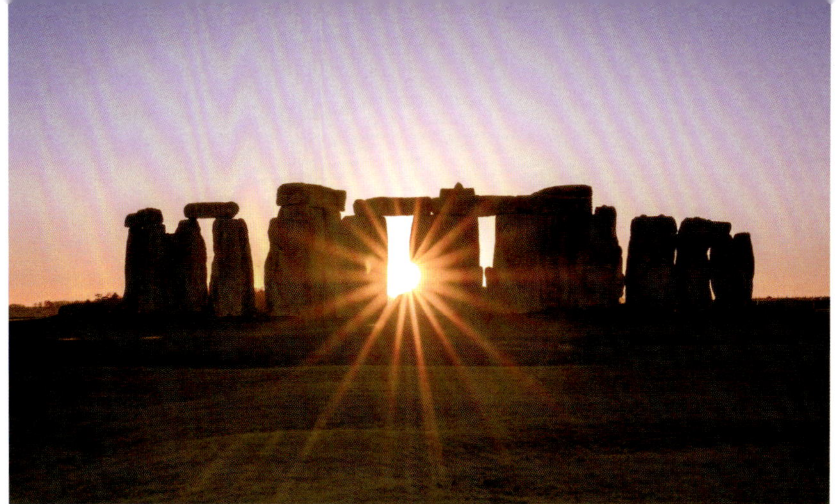

Jo Bourne captured the midwinter sunset at Stonehenge using a Sony A7 IV camera with a 24–105-mm Sony lens (set to 75 mm) at f/20 and ISO 60, with an exposure of 1/13 second.

DECEMBER'S TOPIC: SUPERMOON

We're treated to a particularly big and bright Full Moon on the night of 23/24 December. Our companion world is at its closest point in 2026 – just 356,740 kilometres away – and the Moon appears 14 per cent larger and 30 per cent brighter than the faintest Full Moon.

A Full Moon at the closest point in its orbit (perigee) is called a supermoon. They are not rare: every year we have typically three or four supermoons. But this month's supermoon is the nearest and brightest since 2018, and we'll have to wait until 2034 for a more brilliant display.

The catchy nickname, by the way, wasn't invented by astronomers, but by astrologers who've tried to link supermoons with earthquakes, tsunamis or volcanic eruptions. But the supermoon's extra gravitational pull is actually only 3 per cent more powerful than average – so we can confidently predict it will cause no natural calamities!

DECEMBER'S PICTURE

Today's Druids make it easy for themselves: they hold their Stonehenge ceremonies in the warmth of June. But its builders were actually celebrating chilly Midwinter's Day. Instead of standing inside Stonehenge, watching the summer Sun rising over the outlying Heel Stone, they were stationed at the Heel Stone, observing the pale winter Sun setting through the stone arches – as you can see in Jo Bourne's atmospheric image.

The approach to Stonehenge along a processional track called the Avenue – along with grooves in the bedrock near the Heel Stone – points towards the Midwinter Sunset. And the clinching evidence comes from the remains of great annual feasts held nearby, where thousands of people converged from all over the country. These include the bones of pigs that were killed and eaten there when they were nine months old. Piglets are naturally born in the spring, so the great celebrations at Stonehenge must have been timed for Midwinter. So: forget the Druids, and be at Stonehenge for sunset on 21 December!

DECEMBER'S CALENDAR

SUNDAY	MONDAY	TUESDAY	WEDNESDAY	THURSDAY	FRIDAY	SATURDAY
		1 6.08 am Last Quarter Moon near Jupiter, Mars and Regulus (am)	2	3	4 Moon near Venus and Spica (am)	5 Moon near Venus and Spica (am)
6	7	8	9 0.52 am New Moon	10	11	12
13 Geminids	14 Geminids (am); Geminids	15 Geminids (am)	16	17 5.42 am First Quarter Moon near Saturn	18 Moon near Saturn	19
20	21 Winter Solstice; Moon occults the Pleiades	22	23	24 1.28 am Full Moon; supermoon	25	26
27 Moon near Jupiter, Regulus and Mars	28 Moon near Mars, Regulus and Jupiter	29	30 6.59 pm Last Quarter Moon	31		

SPECIAL EVENTS

- **1 December, am:** to the upper right of the Last Quarter Moon you'll find Mars, Regulus and Jupiter.
- **4 December, am:** the Moon lies to the right of Venus, with Spica between them.
- **5 December, am:** Venus and Spica are above the crescent Moon.
- **Nights of 13/14 and 14/15 December:** after moonset (8 pm) it's an excellent year for observing the slow bright shooting stars of the **Geminid meteor shower**, dust grains shed by the asteroid Phaethon.
- **21 December, 8.50 pm:** the Winter Solstice, when the Sun reaches its lowest point in the sky and the northern hemisphere has the shortest day and the longest night.
- **21 December, 10–11.45 pm:** the Moon passes over the Pleiades, occulting some of its brightest stars (Chart 12a).
- **24 December:** the last of this year's three supermoons is the brightest Full Moon in almost nine years (See Topic).
- **27 December:** the Moon is right next to Regulus, with Jupiter above and Mars to the lower left (Chart 12b).
- **28 December:** above the Moon you'll find Mars, Regulus and Jupiter (Chart 12b).

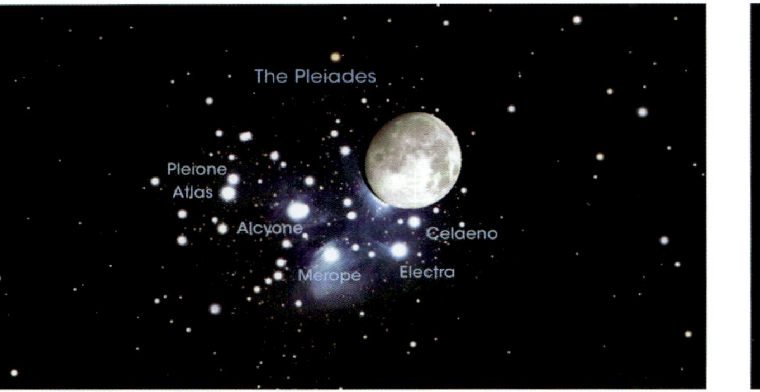

12a 21 December, 11 pm. The Moon occults the Pleiades.

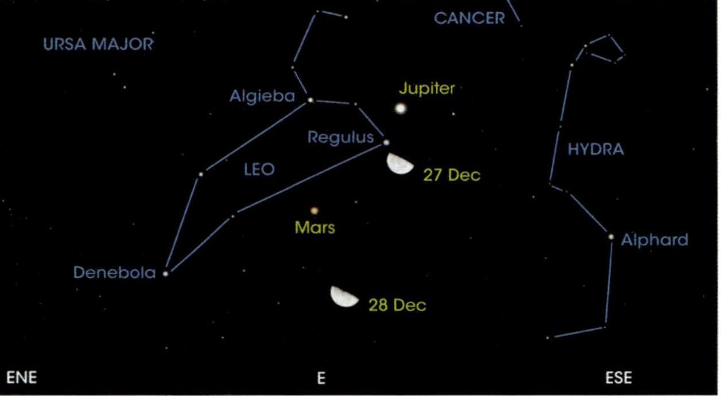

12b 27-28 December, 11 pm. The Moon with Jupiter, Mars and Regulus.

- **Saturn** is the only bright light in a region of dim stars over to the west. It shines at magnitude +0.7 in Cetus and sinks below the horizon around 1 am. The First Quarter Moon joins Saturn on 17 December.
- Setting about 0.30 am, **Neptune** lies in Pisces and glows at only magnitude +7.8.
- **Uranus** (magnitude +5.6) is in Taurus and sets around 6 am.

- **Jupiter** loiters near Regulus, in Leo, all month. Rising about 9 pm, the giant planet shines gloriously at magnitude –2.3. The Moon is nearby on 27 December (Chart 12b).
- At the beginning of December, **Mars** lies just below Jupiter and Regulus, but it moves downwards through Leo during the month. At magnitude +0.2, the Red Planet rises around 10 pm. The Moon lies near Mars on 27 and 28 December (Chart 12b).

- **Venus** is a brilliant Morning Star, rising at 3.45 am and outshining all the stars and the other planets at magnitude –4.8. The crescent Moon forms a striking pair with Venus on the mornings of 4 and 5 December. Through a small telescope you'll see the planet's crescent shape grow gradually wider as the month progresses.

- During the first few days of the month, you can catch **Mercury** well down to the lower left of Venus, some 40 times fainter at magnitude –0.7, and rising at 6 am. But the innermost planet then disappears into the morning twilight glow.

DECEMBER'S PLANET WATCH

Can you see the planets? It's a common question; and the answer is a resounding 'yes!' Some of our cosmic neighbours are the brightest objects in the night sky after the Moon. As they're so close, you can watch them getting up to their antics from night to night. And planetary debris – leftovers from the birth of the Solar System – can light up our skies as glowing comets and the celestial fireworks of a meteor shower.

THE SUN-HUGGERS

Mercury and Venus orbit the Sun more closely than our own planet, so they never seem to stray far from our local star: you can spot them in the west after sunset, or the east before dawn, but never all night long. At *elongation*, the planet is at its greatest separation from the Sun, though – as you can see in the diagram (right) – that's not when the planet is at its brightest. Through a telescope, Mercury and Venus (technically known as the *inferior planets*) show phases like the Moon – from a thin crescent to a full globe – as they orbit the Sun.

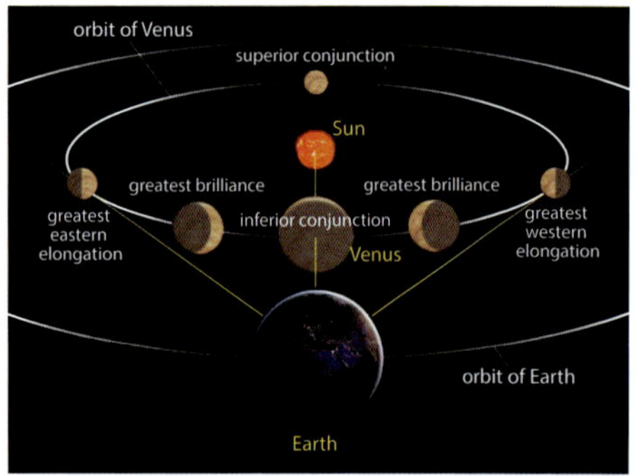

Venus (and Mercury) show phases like the Moon as they orbit the Sun.

Mercury

The innermost planet puts on its best evening performance in February. Mercury is visible low down in early June; but it is lost in the evening twilight at its October elongation. In terms of its morning appearances, Mercury is drowned out by bright twilight at its April elongation, but you can catch the innermost planet again in August (Jupiter lies nearby on **15 August**), and during its November apparition when Mercury is at its best in the morning sky.

Venus

From February to September we can enjoy Venus as the Evening Star, reaching its maximum brilliance on **22 September**. In October, Venus swings between Sun and Earth, to emerge as the Morning Star in November and December (maximum brightness on **27 November**).

Maximum elongations of Mercury in 2026	
Date	Separation
19 February	18° east
4 April	28° west
15 June	25° east
2 August	20° west
12 October	25° east
20 November	20° west

Maximum elongation of Venus in 2026	
Date	Separation
15 August	46° east

WORLDS BEYOND

A planet orbiting the Sun beyond the Earth (known in the jargon as a *superior planet*) can be visible at all times of night, as we look outwards into the Solar System. It lies due south at midnight when the Sun, the Earth and the planet are all in line – a time known as *opposition* (see the diagram, right). Around this time the Earth lies nearest to the planet, although the date of closest approach (and the planet's maximum brightness) may differ by a few days because the planets' orbits are not circular.

Mars

The Red Planet does not come to opposition in 2026. It's lost in the morning twilight until June, and then brightens continuously as Mars heads towards opposition in February 2027. On the morning of **11 October**, the planet passes right in front of Praesepe, and it has a close encounter with Jupiter on **16 November**.

● Where to find Mars	
June–mid-August	Taurus
Mid-August–September	Gemini
October	Cancer
November–December	Leo

Jupiter

The giant planet is nearest to the Earth on **9 January** and at opposition on **10 January**. Lying in Gemini, Jupiter is brilliant in the evening sky until July: Venus is nearby on **9 June**. In August it reappears before dawn in Cancer, before moving into Leo in October, where Jupiter remains for the rest of the year.

Saturn

From January to mid-March, you can

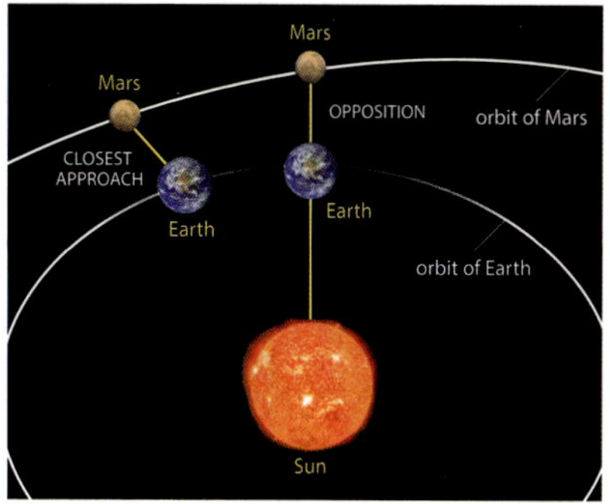

Mars (and the outer planets) line up with the Sun and Earth at opposition, but they are brightest at their point of closest approach.

catch Saturn to the west after sunset, in Pisces. After disappearing behind the Sun, it reappears in Pisces in the morning sky in May. Travelling on into Cetus in September, Saturn is at opposition and at its closest to the Earth on **4 October**.

Uranus

The seventh planet lies in Taurus all year, near the Pleiades: Venus is nearby on **23 April**, and Mars on **4 July**. Until May, Uranus is visible in the evening sky: it re-emerges in the morning sky in July. Uranus reaches opposition on **25 November** and is closest to Earth on **26 November**.

Neptune

Neptune can be seen (though only through binoculars or a telescope) from January to March, and then from June until the end of the year. Saturn is nearby on **20 February**, and Venus on **7 March**. The most distant planet – in Pisces all year – is at opposition and closest to the Earth on **25 September**.

SOLAR ECLIPSES

An annular eclipse of the Sun crosses the coast of Antarctica on **17 February**; it's partial for the rest of Antarctica,

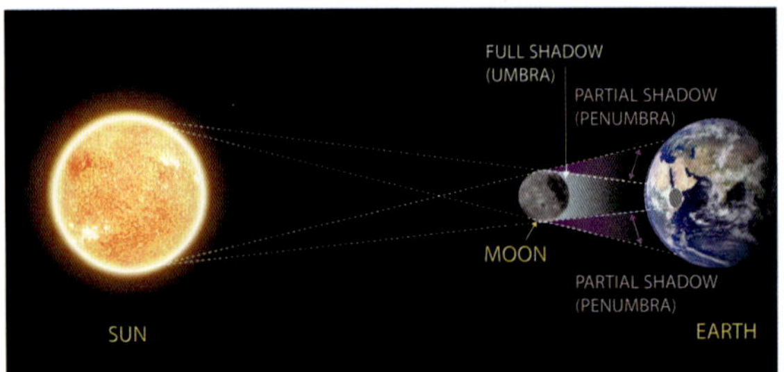

Where the dark central part (the umbra) of the Moon's shadow reaches the Earth, we are treated to a total solar eclipse. If the shadow doesn't quite reach the ground, we see an annular eclipse. People located within the penumbra observe a partial eclipse.

south-east Africa and the southern Indian Ocean, but nothing is visible from Britain and Ireland.

On **12 August**, a total eclipse of the Sun is visible from western Iceland, the north-eastern Atlantic Ocean and northern Spain. Ireland and the UK will witness a very large partial eclipse (up to 98 per cent).

LUNAR ECLIPSES

On **3 March**, the Americas, Asia and Australasia are treated to a total eclipse of the Moon, but nothing is visible from Africa or Europe, including Ireland and the UK.

A large partial lunar eclipse (93 per cent) on **7 September** is visible from Australia, Asia, Africa and Europe. As seen from Ireland and the UK, the Moon rises fully eclipsed and rapidly moves out of the Earth's shadow.

METEOR SHOWERS

Shooting stars, or *meteors,* are tiny specks of interplanetary dust burning up in the Earth's atmosphere. At certain times of year, Earth passes through a stream of debris (usually left by a comet) and we see a *meteor shower.* The meteors appear to emanate from a point in the sky known as the *radiant.* Most showers are known by the constellation in which the radiant lies.

As well as the annual meteor showers listed in the table, this year

Table of annual meteor showers	
Meteor shower	Date of maximum
Quadrantids	3/4 January
Lyrids	22/23 April
Eta Aquarids	6 May (am)
Perseids	12/13 August
Orionids	21/22 October
Leonids	17/18 November
Geminids	13/14 and 14/15 December

there's a chance of celestial fireworks on **9 September**, when the Earth runs into dust particles shed by comet 45P/Honda–Mrkos–Pajdušáková.

COMETS

Comets are dirty snowballs from the outer Solar System. If they fall towards the Sun, its heat evaporates their ices to produce a gaseous head (*coma*) and sometimes dramatic tails. Although some comets are visible to the naked eye, use binoculars to reveal stunning details in the coma and the tail.

Hundreds of comets move round the Sun in small orbits. But many more don't return for thousands or even millions of years. Most comets are now discovered in professional surveys of the sky, but a few are still found by dedicated amateur astronomers. No bright comets are expected in 2026 at the time of writing, but watch out in case a brilliant new comet puts in a surprise appearance!

Here are some of the most popular sights in the night sky, in a season-by-season summary. It doesn't matter if you're a complete beginner, finding your way around the heavens with the unaided eye ◉ or binoculars 🔭; or if you're a seasoned stargazer, with a moderate telescope ⟍. There's something here for everyone.

Each sky sight comes with a brief description, and a guide as to how you can best see it. Many of the most delectable objects are faint, so avoid moonlight when you go out spotting. Most of all, enjoy!

SPRING

Praesepe   ◉ 🔭 ⟍

Constellation: Cancer
Star Chart/Key: March; **5**
Type/Distance: Star cluster; 600 light years
Magnitude: +3.7
A fuzzy patch to the unaided eye; a telescope reveals many of its 1000 stars.

M81 and M82   🔭 ⟍

Constellation: Ursa Major
Star Chart/Key: March; **6**
Type/Distance: Galaxies; 12 million light years
Magnitude: +6.9 (M81); +8.4 (M82)
A pair of interacting galaxies: the spiral M81 appears as an oval blur, and the starburst M82 as a streak of light.

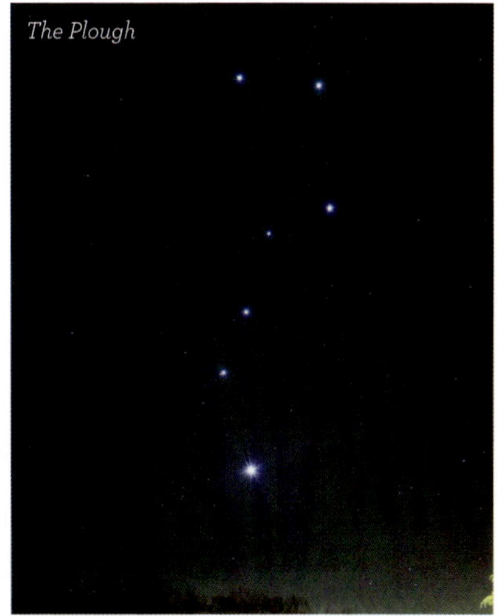
The Plough

The Plough   ◉

Constellation: Ursa Major
Star Chart/Key: April; **7**
Type/Distance: Asterism; 82–123 light years
Magnitude: Stars are roughly magnitude +2
The seven brightest stars of the Great Bear form a large saucepan shape, called 'the Plough'.

Mizar and Alcor   ◉ 🔭 ⟍

Constellation: Ursa Major
Star Chart/Key: April; **8**
Type/Distance: Double star; 83 & 82 light years
Magnitude: +2.3 (Mizar); +4.0 (Alcor)
The sky's classic double star, easily separated by the unaided eye: a telescope reveals Mizar itself is a close double.

Virgo Cluster   🔭 (difficult) ⟍

Constellation: Virgo
Star Chart/Key: May; **9**
Type/Distance: Galaxy cluster; 54 million light years
Magnitude: Galaxies range from magnitude +9.4 downwards
Huge cluster of 2000 galaxies, best seen through moderate to large telescopes.

SUMMER

Antares   ◉ 🔭 ⟍

Constellation: Scorpius
Star Chart/Key: June; **10**
Type/Distance: Red giant; 550 light years
Magnitude: +0.96
Bright red star close to the horizon. You can spot a faint green companion with a telescope.

M13 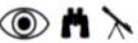

Constellation: Hercules
Star Chart/Key: June; 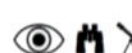
Type/Distance: Star cluster; 23,000 light years
Magnitude: +5.8
A faint blur to the naked eye, this ancient globular cluster is a delight seen through binoculars or a telescope. It boasts nearly a million stars.

Lagoon and Trifid Nebulae 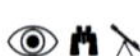

Constellation: Sagittarius
Star Chart/Key: July; 
Type/Distance: Nebulae; 5000 light years
Magnitude: +6.0 (Lagoon); +7.0 (Trifid)
While the Lagoon Nebula is just visible to the unaided eye, you'll need binoculars or a telescope to spot the Trifid. The two are in the same binocular field of view, and present a stunning photo opportunity.

Albireo 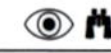

Constellation: Cygnus
Star Chart/Key: August;
Type/Distance: Double star; 360 light years
Magnitude: +3.2 (Albireo A); +5.1 (Albireo B)
Good binoculars reveal Albireo as being double. But you'll need a small telescope to appreciate its full glory. The brighter star appears golden; its companion shines piercing sapphire. It is the most beautiful double star in the sky.

Dumbbell Nebula

Constellation: Vulpecula
Star Chart/Key: August;
Type/Distance: Planetary nebula; 1360 light years
Magnitude: +7.5

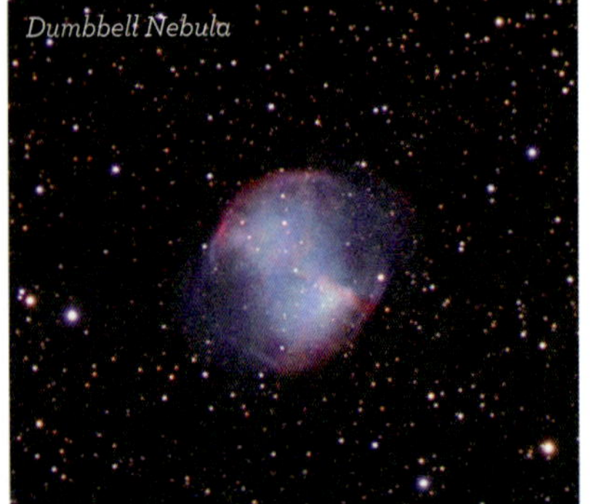

Dumbbell Nebula

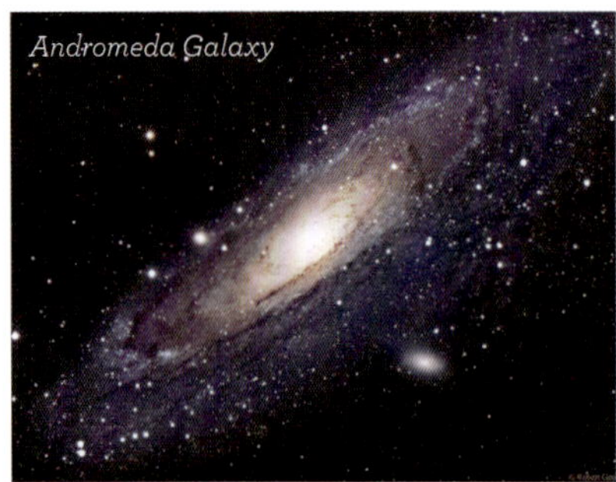

Andromeda Galaxy

Visible through binoculars, and a lovely sight through a small/medium telescope, this dying star has puffed off its atmosphere into space.

AUTUMN

Delta Cephei 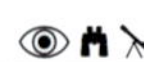

Constellation: Cepheus
Star Chart/Key: September;
Type/Distance: Variable star; 890 light years
Magnitude: +3.5 to +4.4, varying over 5 days 9 hours
The classic variable star, Delta Cephei is chief of the Cepheids – stars that allow us to measure distances in the Universe (their variability time is coupled to their intrinsic luminosity). Visible to the unaided eye, but you'll need binoculars for serious observations.

Andromeda Galaxy 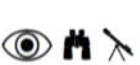

Constellation: Andromeda
Star Chart/Key: October;
Type/Distance: Galaxy; 2.5 million light years
Magnitude: +3.4
The nearest major galaxy to our own, the Andromeda Galaxy is easily visible to the unaided eye in unpolluted skies. Four times the width of the Full Moon, it's a great telescopic object and photographic target.

Mira

Constellation: Cetus
Star Chart/Key: November;
Type/Distance: Variable star; 300 light years
Magnitude: +2 to +10 over 332 days, although maxima and minima may vary.

Nicknamed 'the Wonderful', this distended red giant star is alarmingly variable as it swells and shrinks. At its brightest, Mira is a naked-eye object; binoculars may catch it at minimum; but you need a telescope to monitor this star. Its behaviour is unpredictable, and it's important to keep logging it.

Double Cluster

Constellation: Perseus
Star Chart/Key: November; 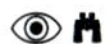
Type/Distance: Star clusters; 7500 light years
Magnitude: +3.7 and +3.8
A lovely sight to the unaided eye, these stunning young star clusters are sensational through binoculars or a small telescope. They're a great photographic target.

Algol

Constellation: Perseus
Star Chart/Key: November; 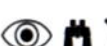
Type/Distance: Variable star; 90 light years
Magnitude: +2.1 to +3.4 over 2 days 21 hours
Like Mira, Algol is a variable star, but not an intrinsic one. It's an 'eclipsing binary' – its brightness falls when a fainter companion star periodically passes in front of the main star. It's easily monitored by the eye, binoculars or a telescope.

WINTER

Pleiades

Constellation: Taurus
Star Chart/Key: December; 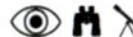
Type/Distance: Star cluster; 440 light years
Magnitude: Stars range from magnitude +2.9 downwards
To the naked eye, most people can see six stars in the cluster, but it can rise to 14 for the keen-sighted. In binoculars or a telescope, they are a must-see. Astronomers have observed 1000 stars in the Pleiades.

Orion Nebula

Constellation: Orion
Star Chart/Key: January; 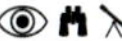
Type/Distance: Nebula; 1340 light years
Magnitude: +4.0

A striking sight even to the unaided eye, the Orion Nebula – a star-forming region 24 light years across – hangs just below Orion's Belt. Through binoculars or a small telescope, it is staggering. A photographic must!

Betelgeuse

Constellation: Orion
Star Chart/Key: January; 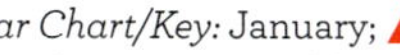
Type/Distance: Variable star; 720 light years
Magnitude: 0.0 to +1.6
Even with the unaided eye, you can see that Betelgeuse is slightly variable over months, as the red giant star billows in and out.

M35

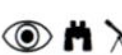

Constellation: Gemini
Star Chart/Key: February; 3
Type/Distance: Star cluster; 2800 light years
Magnitude: +5.3
Just visible to the unaided eye, this cluster of around 2000 stars is a lovely sight through a small telescope.

Sirius

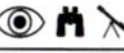

Constellation: Canis Major
Star Chart/Key: February; 4
Type/Distance: Double star; 8.6 light years
Magnitude: –1.5
You can't miss the Dog Star. It's the brightest star in the sky! But you'll need a 150-mm reflecting telescope (preferably bigger) to pick out its +8.4 magnitude companion – a white dwarf nicknamed 'the Pup'.

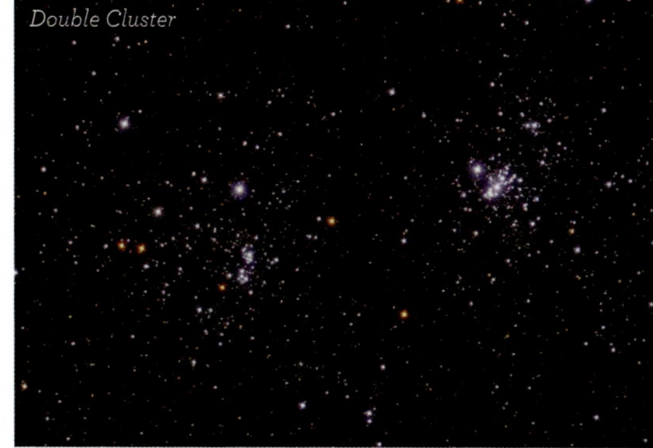

Telescopes have never been better value, and many people would like to have one of their own, even if they don't plan to go full tilt into the hobby. These days there are more options than ever before, but the extensive choice means making a decision – which is the best type to go for?

I'll assume that you already have binoculars, which are ideal for general stargazing and will show a wide variety of objects from the craters of the Moon to the larger star clusters and nebulae and a handful of galaxies. But there comes a point where the limited magnification of binoculars just isn't enough. In addition, as the magnification goes up, the harder it is to keep the instrument still, and binoculars are difficult to keep steady when observing objects high in the sky. This highlights the crucial point that the mounting of whatever instrument you choose is just as important as the optical quality – even at the most basic level.

To narrow down your choice, consider your local circumstances – basically, whether you have a lot of light pollution that restricts what you are most likely to be able to view – and whether you want to stick to visual observing, or are keen to use imaging techniques. Even from brightly lit areas you can now use imaging to get amazing views of both bright solar-system objects and the fainter deep-sky objects such as nebulae and galaxies, although at greater but not excessive cost.

If your night sky is badly light-polluted, for visual observing you are mostly restricted to the Moon and planets. A few deep-sky objects are visible, such as star clusters, but the fainter objects are generally invisible against the background or at best washed out. But don't let that put you off – some leading amateur astronomers do their work from such locations. In more rural sites you have a wider choice of objects to observe and your choice of instrument is greater.

There are basically two main types of telescope – refractors, with a lens at the top of the tube, and reflectors, which use a mirror at the bottom of the tube to focus the light. Refractors are generally more robust and maintenance-free, but the lower-priced achromatic models suffer from false colour, which shows up as a blue haze around bright objects. They are also costly in large sizes. Reflectors give a colour-free image, but the mirrors are delicate and can get out of line, so they need more care. However, there's no restriction on mirror size and quite large reflectors are not particularly expensive. The diameter of the lens or mirror is referred to as the aperture.

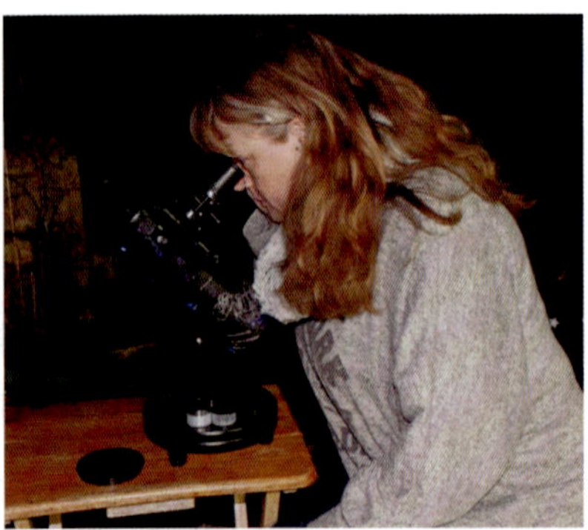

Mary McIntyre is an experienced observer and astrophotographer with several advanced telescopes, but still enjoys viewing with her tabletop Dobsonian.

Both types are available at the low-budget end of the market, costing under £100. If you just want a simple telescope for occasional use, the classic 60-mm refractor on tripod will give reasonable views of the Moon and major planets up to a magnification of about 70, but the mountings are notoriously unsteady because of the small central pivot, and for a more enjoyable experience I would recommend paying more. A more stable view is provided by the tabletop reflecting telescopes, sitting on a turntable system known as a Dobsonian mount. These are basic Newtonian-type reflectors, with a main concave mirror at the bottom of the tube and a smaller flat mirror near the top to reflect the light to an interchangeable eyepiece. As with all astronomical telescopes, each eyepiece gives a different magnification.

Between £100 and £250 you can get a range of reflectors in the tabletop range, some with parabolic rather than spherical concave mirrors, so they can give a wider field of sharp view, and some with collapsible tubes for travelling. Or there are achromatic refractors up to about 90-mm aperture and on more sturdy mountings than the very low-priced models.

Any mounting that provides simple up-and-down and side-to-side movements is known as an altazimuth type, but many astronomical telescopes are on equatorial mountings. These need to be aligned with one axis parallel to Earth's axis before use, so as to track objects with a single motion, and also require a heavy counterweight. Some budget telescopes are on equatorial mountings, but I feel that they are more trouble than they are worth for beginners and I would recommend an altazimuth mount.

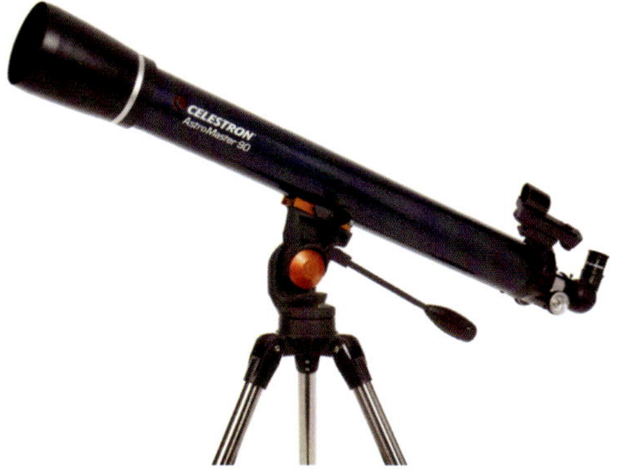

The Celestron Astromaster 90AZ is an f/11 achromatic refractor which offers magnifications up to 100 and is mounted on a dovetail so that it can be used on different mountings.

The simplicity and sturdiness of a Dobsonian mount mean it has a lot going for it, and Dobsonian-type telescopes are available in sizes up to the largest that amateurs are likely to need. Anything over 150-mm aperture becomes rather cumbersome, even if it has a collapsible tube so, although they are transportable in a car, they do take up a lot of space which you need to take into account.

A word about focal ratio, usually written as f/5 or f/9, for example. This is the focal length of the mirror or lens – in effect, the length of the tube – divided by the aperture. For a given aperture, the lower the number the shorter the tube length, but the less likely it is to give good results at high magnifications, particularly with budget instruments. If you want good views of the Moon and planets, particularly in light-polluted areas, the longer focal ratios are generally recommended. Short focal ratios are best suited to more rural areas as they have the potential to give brighter and wider-field views of faint objects. But these are just guidelines, and any telescope is better than no telescope if used to its best advantage.

VERSATILITY

If you feel that you may want to develop your interest in astronomy, bear in mind that some of the lowest-cost instruments have the limitation that the telescopes can't be detached from their mountings. This severely restricts their versatility as you can't then put the telescope itself (known in the jargon as an OTA, for 'optical tube assembly') on a better mounting. Beyond the beginner-range, however, astronomical telescopes and mountings use the dovetail system of connection between the two. The 45-mm Vixen dovetail is the standard, although larger sizes are used for advanced instruments. Some tabletop Dobsonians are attached by standard dovetails, increasing their usefulness.

Another way you might want to develop your interest is to use your telescope for photography. The simplest

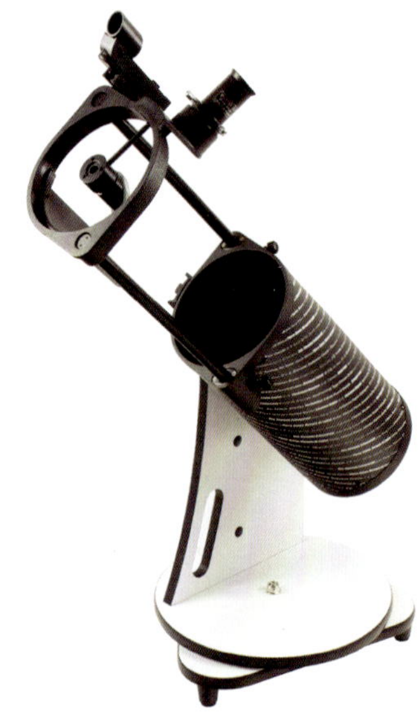

This Sky-Watcher Heritage 130P flextube tabletop f/5 reflector can give excellent views of deep-sky objects and has enough aperture to show detail on the planets with its parabolic mirror.

is to point your phone camera into the eyepiece when observing the Moon and take a snap, which can often give surprisingly good results although lining everything up is pot luck. The better way is to replace the eyepiece with a camera body, so as to use the telescope as a telephoto lens. This is where problems usually start.

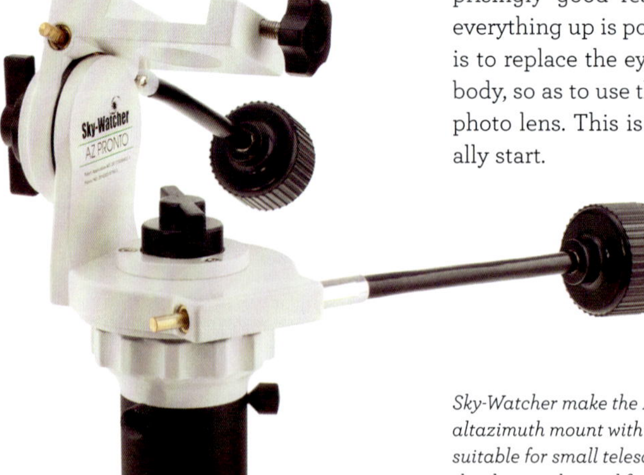

Sky-Watcher make the Pronto AZ, an inexpensive altazimuth mount with slow-motion controls suitable for small telescopes up to 3 kg in weight that have a dovetail fitting.

Many cameras are so heavy that they will overbalance the instrument, so you may need to improvise a counterweight. More of a problem is that the limited focusing range of the telescope may not allow for the back focus of the camera – the distance between the lens mount and the sensor inside the camera – so you can never focus the image. This is more of a problem with reflectors than refractors, which usually have a star diagonal in their viewing system that gives a better viewing angle. Replacing this with a short extension tube, available from telescope suppliers, usually does the trick.

So if you feel that you may want greater versatility when choosing your first telescope, aim for one that is fixed to its mounting with a dovetail. Altazimuth mounts with dovetail saddles are available that will take instruments up to about 5 kilograms in weight, providing a convenient mounting for a small instrument. You can choose the items separately or as a bundle.

Some altazimuth mounts are available with a GoTo function. These require a power supply and can find objects in the sky for you, once aligned on a few chosen stars. Older versions used separate handsets that contained the computer control system, but these have been largely replaced by systems that link to an app on your phone or tablet. Some GoTo altazimuth mounts can be converted into equatorial mounts by adding what's called a wedge, which allows for long-exposure photography.

By buying separate items, your first telescope could remain useful to you for many years as a small portable instrument for travel or for a quick look, with the mounting being able

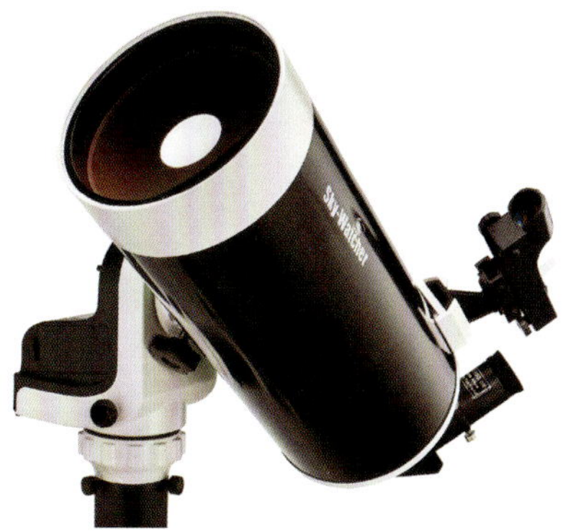

The Maksutov–Cassegrain optical design of this Sky-Watcher SkyMax-127 AZ GTi provides a long focal ratio of f/11.8 in a compact tube, best suited to planetary observing. The AZ GTi altazimuth mount will find and track a wide range of objects from an app and is available with a range of alternative telescopes.

to take a higher-grade instrument for more detailed views or even imaging. Apochromatic or ED refractors, for example, will provide almost colour-free images compared with the cheaper achromatic models.

Alternatively, to complement a purely visual low-cost instrument for the occasional look at what's in the sky, you could also buy one of the increasingly good value smart telescopes that will align itself and send images of deep-sky objects to your smartphone or tablet. These can even take decent images from light-polluted areas with the use of specialised filters.

One last piece of advice. Specialist telescope suppliers will always be happy to advise you, by email or by phone, on whether your choice of instrument will do what you want, such as being adaptable for photography. Their websites often contain a wealth of information and customer reviews that will help you to make the right choice.

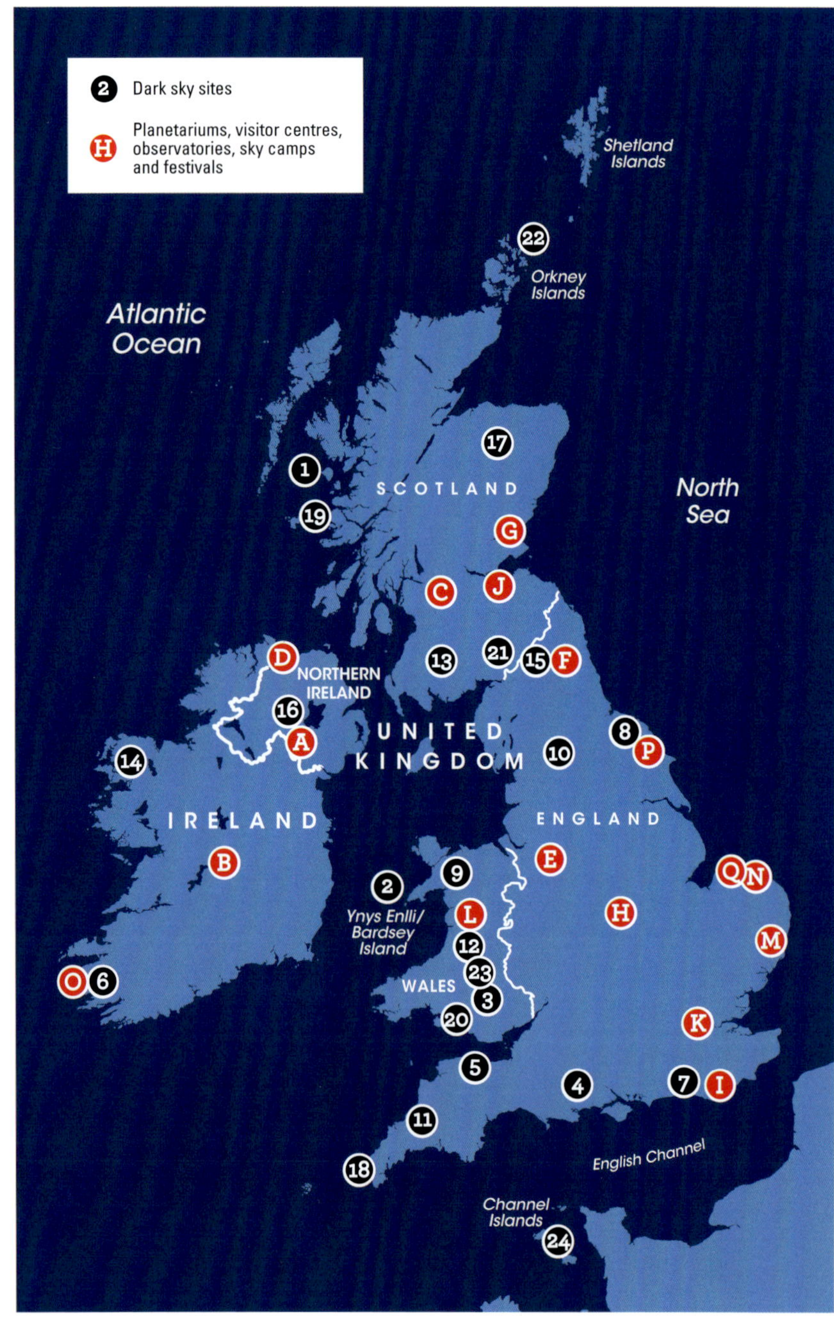
Dark sky sites
Planetariums, visitor centres, observatories, sky camps and festivals
Shetland Islands
Atlantic Ocean
Orkney Islands
SCOTLAND
North Sea
NORTHERN IRELAND
UNITED KINGDOM
IRELAND
ENGLAND
Ynys Enlli/ Bardsey Island
WALES
English Channel
Channel Islands

Take a break from your backyard stargazing! View the pristine heavens from a protected dark-sky site, visit a major observatory or planetarium, or join in the fun at a star camp or astronomical event.

DARK-SKY SITES

The International Dark-Sky Association, based in Tucson, Arizona, checks out the darkest places in the world. In these islands, 24 places are internationally recognised for their unsullied view of the heavens or their commitment to combating light pollution. Many are in National Landscapes, formerly Areas of Outstanding Natural Beauty.

DARK SKY SANCTUARIES

Noted for exceptionally dark nights and a nocturnal environment protected for its scientific and cultural heritage, a Dark Sky Sanctuary is often in a remote location.

1. Isle of Rum

Designated: 2024 • *Area:* 105 sq km
Nearest town: Mallaig
Accessible only to walkers and cyclists, the Isle of Rum is home to 120,000 pairs of Manx shearwaters as well as golden eagles and white-tailed sea eagles. It has only 40 inhabitants and no public lighting.

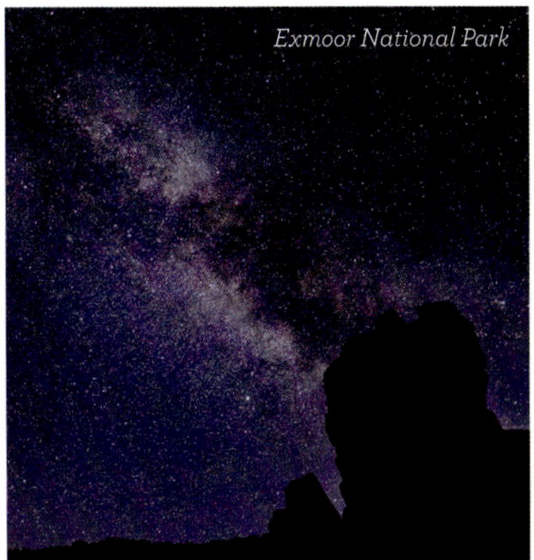

2. Ynys Enlli/Bardsey Island

Designated: 2023 • *Area:* 1.8 sq km
Nearest town: Pwllheli
Europe's first designated Dark Sky Sanctuary is home to a major bird observatory, with an emphasis on nocturnal species. It lies 3 km off the Welsh coast, and is accessible only by small boat during good weather.

DARK SKY RESERVES

Dark Sky Reserves are generally large, and consist of a very dark core region surrounded by a peripheral area where lighting is minimal. Eight out of a total of 19 worldwide are in Britain and Ireland.

3. Bannau Brycheiniog/
Brecon Beacons National Park

Designated: 2013 • *Area:* 1347 sq km
Nearest towns: Brecon, Merthyr Tydfil
Situated in the mountains of South Wales, where sheep outnumber humans 30 to 1, this Reserve is home to 33,000 people – yet lighting is controlled so that the core zone has some of the darkest skies in Wales.

4. Cranborne Chase

Designated: 2019 • *Area:* 981 sq km
Nearest towns: Salisbury, Shaftesbury
Not far from Stonehenge, Cranborne Chase is a chalk plateau rising to 277 m at Win Green. This National Landscape has commanding views to the west and to some extent to the north.

5. Exmoor National Park

Designated: 2011 • *Area:* 693 sq km
Nearest towns: Barnstaple, Minehead, Taunton
The first Dark Sky Reserve site to be designated in these islands, Exmoor is a moorland area with much preserved history and many monuments within the 81-sq-km core zone.

6. Kerry

Designated: 2014 • *Area:* 700 sq km
Nearest towns: Kenmare, Waterville
Its location between the Kerry Mountains and the Atlantic Ocean gives this Dark Sky Reserve natural protection from light pollution, yet it's readily accessible from the Wild Atlantic Way, the stunning tourist route that runs through the reserve.

7. Moore's Reserve (South Downs)

Designated: 2016 • *Area:* 1627 sq km
Nearest towns: Brighton, Portsmouth
Named after the British astronomy populariser Sir Patrick Moore (1923–2012) who lived nearby, this reserve is sandwiched between London and the busy seaside resorts of Brighton and Worthing – yet it retains remarkably dark skies.

8. North York Moors National Park

Designated: 2020 • *Area:* 1440 sq km
Nearest towns: Scarborough, Whitby
Despite its proximity to the busy tourist destinations of Whitby and Scarborough, the North York Moors is a largely deserted expanse of heather and bog moorland, with wide views of the night sky.

9. Eryri/Snowdonia National Park

Designated: 2015 • *Area:* 2132 sq km
Nearest towns: Harlech, Porthmadog

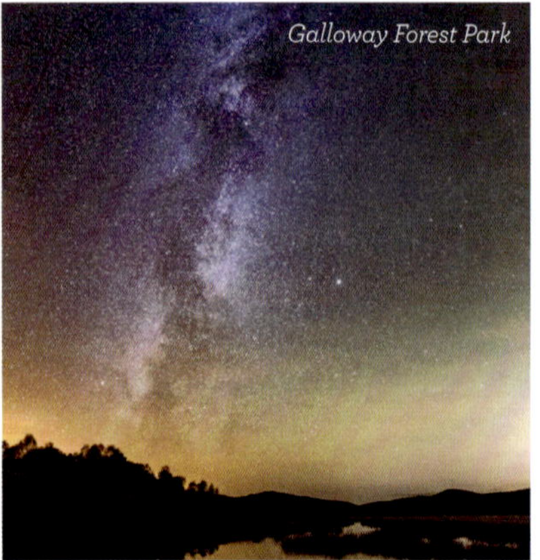

The national park encompasses around 10 per cent of the total land area of Wales, and the darkest skies are to be seen from the rugged central area around Mount Snowdon (1085 m).

10. Yorkshire Dales National Park

Designated: 2020 • *Area:* 2180 sq km
Nearest towns: Hawes, Kirby Lonsdale
The largest dark-sky site in Britain and Ireland, the Yorkshire Dales National Park boasts impressive waterfalls and caves, and brilliant night skies within reach of major cities like Leeds and Manchester.

DARK SKY PARKS

Smaller regions with exceptionally low light pollution are designated Dark Sky Parks. Currently numbering eight in Britain and Ireland, more are being added every year.

11. Bodmin Moor Dark Sky Landscape

Designated: 2017 • *Area:* 208 sq km
Nearest towns: Bodmin, Launceston, Liskeard
This remote, rugged area of granite moorland in north-east Cornwall is a working agricultural landscape, protecting it from large-scale development that would threaten dark skies.

12. Elan Valley Estate

Designated: 2015 • *Area:* 180 sq km
Nearest towns: Aberystwyth, Rhayader
The city of Birmingham purchased this Welsh valley in 1892 to construct reservoirs that would provide a regular supply of pure water. Views of the starry night over the reservoirs are particularly impressive.

13. Galloway Forest Park

Designated: 2009 • *Area:* 780 sq km
Nearest towns: Girvan, Newton Stewart
Galloway Forest Park is the largest forest park in the UK, and 20 per cent has been set aside as a core area with no illumination allowed. It's a mecca not just for astronomers but for nocturnal wildlife whose lives are often disrupted elsewhere by light pollution.

14. Mayo Dark Sky Park

Designated: 2016 • *Area:* 150 sq km
Nearest towns: Ballina, Castlebar
One of the largest expanses of peat landscape in Europe, this region supports a unique diversity of bog-dwelling species. Unsuitable for agriculture, and adjoining the Atlantic Ocean, the park should enjoy pristine skies far into the future.

15. Northumberland National Park and Kielder Water & Forest Park

Designated: 2013 • *Area:* 1592 sq km
Nearest towns: Jedburgh, Rothbury
Near Hadrian's Wall, built to keep the Picts from Roman Britain, this Dark Sky Park was designated as a bulwark against light pollution invading the darkness of northern England. It contains the largest reservoir and most extensive forest in northern Europe.

16. OM Dark Sky Park & Observatory

Designated: 2020 • *Area:* 15 sq km
Nearest towns: Cookstown, Magherafelt
The first dark-sky place accredited in Northern Ireland, OM is set among rolling hills and is centred on the Bronze Age site of Beaghmore Stone Circles.

17. Tomintoul and Glenlivet, Cairngorms

Designated: 2018 • *Area:* 230 sq km
Nearest towns: Dufftown, Grantown-on-Spey
Containing the dramatic landscape of the Cairngorm Mountains, this Dark Sky Park is home to Britain's only herd of wild reindeer. And, if the weather is cloudy, the Park also contains the Glenlivet whisky distillery!

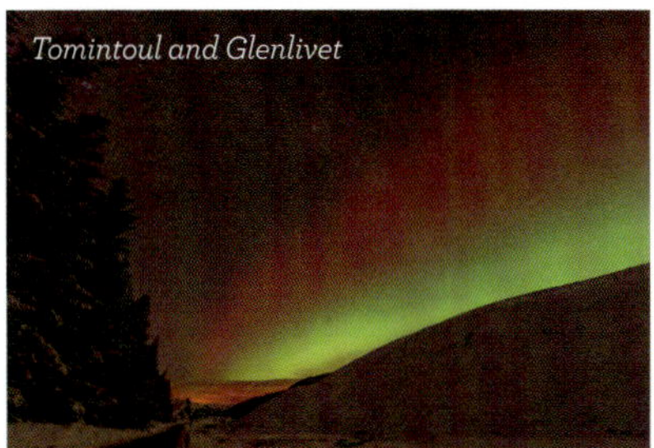
Tomintoul and Glenlivet

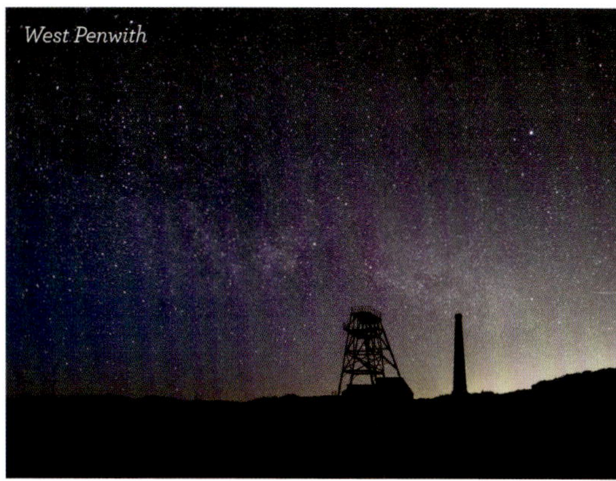

18. West Penwith

Designated: 2021 • *Area:* 136 sq km
Nearest towns: Penzance, St Ives
The very western tip of Cornwall, stretching down to Land's End, West Penwith is a wild landscape, with stunning sea views and ancient archaeological remains – some thought to be astronomically aligned.

DARK SKY COMMUNITIES

A town, village or complete island that's actively fighting light pollution can be designated a Dark Sky Community, with five listed in Britain and Ireland so far, plus one in the Channel Islands.

19. Coll

Designated: 2013 • *Area:* 77 sq km
The Scottish island of Coll is home to just 200 permanent residents, plus myriad birds in its extensive nature reserve. The island has adopted a light-management plan to ensure its skies remain dark.

20. Gower National Landscape

Designated: 2025 • *Area:* 186 sq km
A wild peninsula near Swansea, Gower is known for its stunning landscapes. Now, all its 1641 streetlights have been replaced with dark-sky-approved lamps to minimise light pollution.

21. Moffat

Designated: 2016 • *Area:* 147 sq km
This former spa town is a tourist base for southern Scotland. It has strict outdoor lighting policies to reduce light pollution in its hinterland.

22. North Ronaldsay

Designated: 2021 • *Area:* 7 sq km
The northernmost island in Orkney. Many visitors come to its bird observatory, and the island's appeal extends to astronomers with its Dark Sky Community designation.

23. Presteigne & Norton Dark Sky Community

Designated: 2023 • *Area:* 40 sq km
Two neighbouring towns in Wales combined forces to severely curtail night-time lighting in an area that's home to the Spaceguard Centre observatory and to colonies of light-sensitive bats.

24. Sark

Designated: 2011 • *Area:* 5 sq km
One of the Channel Islands, Sark was Europe's first Dark Sky Community. Its pitch-black skies are largely due to the island's ban on public lighting and motor vehicles apart from tractors.

PLANETARIUMS, VISITOR CENTRES AND OBSERVATORIES

For a great day out under the cloudless sky of a planetarium, a tour of a world-beating observatory or an evening observing the stars, here are some leading locations. Many venues combine a planetarium (⌂), visitor centre (Ⓥ) and observatory (◓).

A. Armagh Observatory and Planetarium

Armagh has the longest-running planetarium and the second-oldest observatory in the UK. Combine a state-of-the-art digital planetarium show with a tour of the observatory's ancient instruments (book in advance for the latter).

B. Birr Castle Demesne 

Marvel at the well-preserved remains of the Leviathan of Parsonstown, the largest telescope in the world from 1845 to 1917. The Science Centre chronicles how the Third Earl of Rosse created his great instrument, and discovered the spiral shape of galaxies.

C. Glasgow Science Centre Planetarium 

One of Scotland's most popular visitor attractions, Glasgow Science Centre features a 15-m diameter planetarium. After viewing the night sky in comfort, you can be awed by the giant screen of the Centre's IMAX cinema.

D. Inishowen Planetarium

Part of the Inishowen Maritime Museum, the planetarium is picturesquely set beside Lough Foyle. As well as putting on astronomy shows, the Maritime Museum uses the planetarium dome to project immersive ocean experiences.

E. Jodrell Bank Discovery Centre 

Grab a close-up view of the Lovell Telescope, the great radio dish that tracked rockets in the early days of the space race, downloaded pictures from the Moon and now investigates the secrets of pulsars. Other displays include exhibitions, talks and interactive activities.

F. Kielder Observatory

The Kielder Observatory, sited under some of the darkest skies in England, has a variety of telescopes for visual observing and astrophotography. It hosts some 700 events per year.

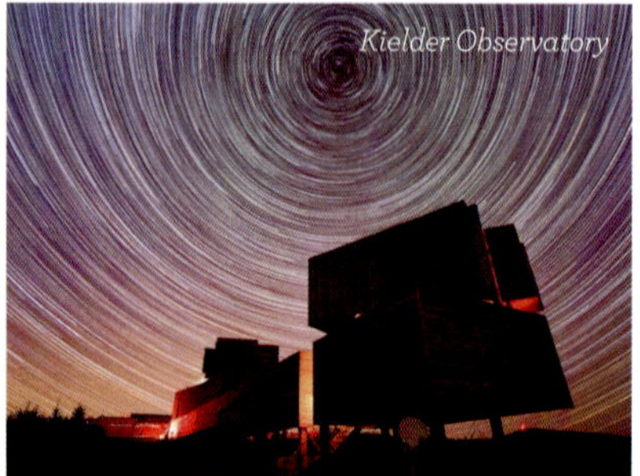

G. Mills Observatory 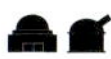

On a hill above Dundee, the Mills Observatory was the UK's first purpose-built public observatory, and it carries on opening its doors to the public every clear weeknight. There's also a small planetarium and a gift shop.

H. National Space Centre 

Located on the outskirts of Leicester, this visitor centre focuses on the history – and future – of the UK in space exploration. Explore the wider Universe in the 192-seat Sir Patrick Moore planetarium.

I. Observatory Science Centre 

On a hillside at Herstmonceux in Sussex, the Observatory Science Centre is located within a cluster of green domes that once housed the telescopes of the Royal Greenwich Observatory. As well as fascinating exhibits, the observatory holds regular stargazing evenings.

J. Royal Observatory, Edinburgh 

Scotland's premier observatory no longer makes professional observations: its astronomers now build and use large telescopes on Hawai'i and in Chile. But the Edinburgh site is still active, with a visitor centre, regular astronomical talks and public stargazing evenings.

K. Royal Observatory and Peter Harrison Planetarium 

The home of British astronomy, the Royal Observatory at Greenwich is a fascinating museum of astronomy and timekeeping: stand on the Meridian Line, with a foot in each hemisphere! The planetarium hosts a variety of astronomical shows.

STAR CAMPS AND FESTIVALS

Whether you're a beginner or a seasoned sky-watcher, enjoy a weekend under the stars with fellow astronomers at these annual star camps and festivals. Also, check out events at the countries' dark-sky sites (pages 91–93) as many of them host Dark Skies Festivals. And there's a comprehensive list of sky-watching activities at gostargazing.co.uk.

L. Cambrian Mountains Stargazing Weekends

Staylittle, Powys • Most months
Stay in a deluxe wigwam cabin under the dark skies of Mid Wales, for a weekend of stargazing, plus tuition in astronomy and astrophotography.

M. Haw Wood Farm Astronomers' Week

Saxmundham, Suffolk • October
Astronomers literally pitch up at Haw Wood for a small-scale, friendly week of stargazing, organised by local astronomical societies.

N. Kelling Heath Star Party

Kelling Heath, Norfolk • March and September
Claiming to be the largest star party in the UK, this biennial event fills three fields with astronomers, tents, trade stands and top-end amateur telescopes.

O. Skellig Star Party

Ballinskelligs, County Kerry • August
Held by the shores of the Atlantic Ocean, Ireland's premier star party features talks by leading astronomers as well as night-time observing.

P. StarFest

Dalby Forest, North Yorkshire • August
This event attracts astronomers from all over the country for a weekend of events that can include rocket-building, talks, an astronomical pub quiz and – of course! – observing the sky.

Q. WinterFest Astro Star Party

Kelling Heath, Norfolk • November
Organised by the Birmingham Astronomical Society around the time of New Moon, to give its members some respite from the city's lights, WinterFest is open to all who want to view the glittering winter stars.

THE AUTHOR

Professor **Nigel Henbest** is an award-winning British science writer, specialising in astronomy and space. After research in radio astronomy at Cambridge, he became a consultant to both the Royal Greenwich Observatory and *New Scientist* magazine.

Originally with Heather Couper, Nigel has been writing Philip's *Stargazing* since 2005. The author of 50 other books, Nigel also co-founded a TV production company where he produced acclaimed programmes and international series on astronomy and space.

Now an Adjunct Professor at the University of North Carolina, Wilmington, Nigel is a Fellow of the Royal Astronomical Society and a Future Astronaut with Virgin Galactic; asteroid 3795 is named 'Nigel' in his honour.

Married with two stepdaughters, Nigel lives in Buckinghamshire and North Carolina.

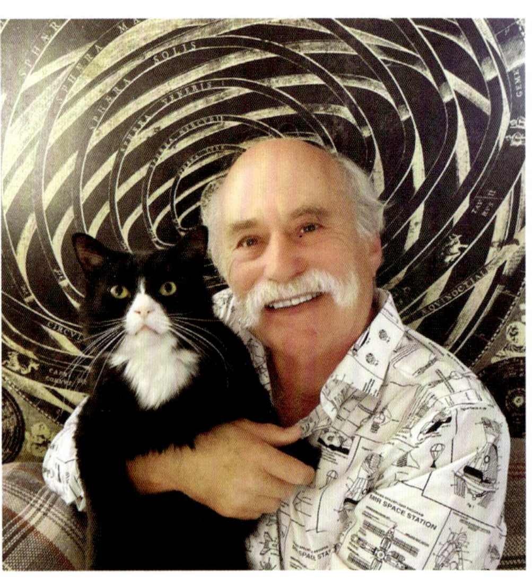

ACKNOWLEDGEMENTS

PHOTOGRAPHS
Front cover: Stonehenge Dronescapes. **Aboyne Photographics/Alamy Stock Photo:** 93 below; **Stuart Atkinson:** 35; **Jo Bourne:** 77; **Celestron:** 87; **Chrispo/Alamy Stock Photo:** 91; **Gary Cook/Alamy Stock Photo:** 92; **Neil Corke:** 29; **Josh Dury Photo-Media:** 53; **Robert Gendler/NASA:** 84 above; **First Light Optics:** 88, 89; **Alexandra Hart:** 23; **Bridget Henbest:** 96; **Nigel Henbest:** 7; **Simon Hudson:** 1; **Peter Jenkins FRAS:** 17; **JPL-Caltech/NASA:** 37; **Pete Lawrence:** 11; **Mary McIntyre:** 71, 86; **Keith Mason/Alamy Stock Photo:** 93 above; **Dan Monk, Kielder Observatory:** 94; **David Moug via Wikipedia:** 36; **Nblahmd (CC by SA-4.0, https://commons.wikimedia.org/wiki/File:Annular_eclipse.jpg):** 18; **NPL/NASA:** 60, 66; **D. Peach:** 65; **Martin Ratcliffe Photography:** 47; **Aaron Ruy Musa (CC by SA-4.0, https://commons.wikimedia.org/wiki/File:Between_Ancient_and_Modern_Lights.jpg):** 30; **Robin Scagell:** 6, 84 below; **Judy Schmidt (CC by 2.0, https://commons.wikimedia.org/wiki/File:Infrared_Rho_Ophiuchi_Complex.jpg):** 48; **Peter Shah:** 85; **UCLA/NASA:** 2–3; **University of Arizona/NASA:** 31; **VegaStar Carpentier/NASA:** 83; **Sara Wager, swagastro.com:** 40; **Andy Weller – Dark Matters:** 59.

ARTWORKS
Star Charts: Wil Tirion/Philip's with extra annotation by Philip's.
Special Event Charts: Nigel Henbest/Stellarium (www.stellarium.org).
Pages 80–82: Chris Bell.
Page 90: Philip's